Springer Series in

MATERIALS SCIENCE 66

Springer-Verlag Berlin Heidelberg GmbH

Springer Series in
MATERIALS SCIENCE

Editors: R. Hull R. M. Osgood, Jr. J. Parisi H. Warlimont

The Springer Series in Materials Science covers the complete spectrum of materials physics, including fundamental principles, physical properties, materials theory and design. Recognizing the increasing importance of materials science in future device technologies, the book titles in this series reflect the state-of-the-art in understanding and controlling the structure and properties of all important classes of materials.

61 **Fatigue in Ferroelectric Ceramics and Related Issues**
By D.C. Lupascu

62 **Epitaxy**
Physical Principles and Technical Implementation
By M.A. Herman, W. Richter, and H. Sitter

63 **Fundamentals of Ion-Irradiated Polymers**
By D. Fink

64 **Morphology Control of Materials and Nanoparticles**
Advanced Materials Processing and Characterization
Editors: Y. Waseda and A. Muramatsu

65 **Transport Processes in Ion-Irradiated Polymers**
By D. Fink

66 **Multiphased Ceramic Materials**
Processing and Potential
Editors: W.-H. Tuan and J.-K. Guo

67 **Nondestructive Materials Characterization**
With Applications to Aerospace Materials
Editors: N.G.H. Meyendorf, P.B. Nagy, and S.I. Rokhlin

68 **Diffraction Analysis of the Microstructure of Materials**
Editors: E.J. Mittemeijer and P. Scardi

69 **Chemical-Mechanical Planarization of Semiconductor Materials**
Editor: M.R. Oliver

70 **Applications of the Isotopic Effect in Solids**
By V.G. Plekhanov

71 **Dissipative Phenomena in Condensed Matter**
Some Applications
By S. Dattagupta and S. Puri

72 **Predictive Simulation of Semiconductor Processing**
Status and Challenges
Editors: J. Dabrowski and E.R. Weber

73 **SiC Power Materials**
Devices and Applications
Editor: Z.C. Feng

Series homepage – springer.de

Volumes 10–60 are listed at the end of the book.

W.-H. Tuan J.-K. Guo (Eds.)

Multiphased Ceramic Materials

Processing and Potential

With 88 Figures and 17 Tables

Springer

Professor Wei-Hsing Tuan
National Taiwan University, Institiute of Materials Science and Engineering
Taipei, Taiwan 106, ROC
E-mail: tuan@ccms.ntu.edu.tw

Professor Jing-Kun Guo
Shanghai Institute of Ceramics, Chinese Academy of Science
The State Key Laboratory of High Performance Ceramics and Superfine Structure
1295 Ding-xi Road, Shanghai, 200050, China

Series Editors:

Professor Robert Hull
University of Virginia, Dept. of Materials Science and Engineering, Thornton Hall
Charlottesville, VA 22903-2442, USA

Professor R. M. Osgood, Jr.
Microelectronics Science Laboratory, Department of Electrical Engineering
Columbia University, Seeley W. Mudd Building, New York, NY 10027, USA

Professor Jürgen Parisi
Universität Oldenburg, Fachbereich Physik, Abt. Energie- und Halbleiterforschung
Carl-von-Ossietzky-Strasse 9-11, 26129 Oldenburg, Germany

ISSN 0933-033X

ISBN 978-3-642-62278-6 ISBN 978-3-642-18752-0 (eBook)
DOI 10.1007/978-3-642-18752-0

Multiphased ceramic materials: processing and potential / W.H. Tuans, J.K. Guo (eds.) p.cm. – (Springer series in materials sciences, ISSN 0933-033X;v.66) Includes bibliographical references and index.
1. Ceramic materials. I. Guo, Jingkun. II. Tuan, W.-H. (Wei-Hsing), 1958- III.
Series. TA455.C4M85 2004 620.1'4–dc22

Bibliographic information published by Die Deutsche Bibliothek Die Deutsche Bibliothek lists this publication in the Deutsche Nationalbibliografie; detailed bibliographic data is available in the Internet at <http://dnb.ddb.de>

springeronline.com

Softcover reprint of the hardcover 1st edition 2004
Originally published by Springer-Verlag Berlin Heidelberg New York in 2004

Typesetting by the editors
Cover concept: eStudio Calamar Steinen
Cover production: *design & production* GmbH, Heidelberg

Printed on acid-free paper 57/3141/ts 5 4 3 2 1 0

Preface

In many fields of thermomechanical applications, such as advanced gas turbine, thermal protection of space planes, oil and coal electric generation, advanced brake devices, etc. high performance materials are needed. Lacking of toughness and reliability, monolithic ceramics had reached their limits and multi-phase materials are expected as the most promising candidates. Apart from the improvement on the thermomechanical performance, more than one function, such as magnetic, electrical and optical functions, can be incorporated simultaneously into the multi-phase material.

However, several key problems remained to be solved in the near future. These problems include microstructure complexity, process challenge, machining ability and the most important cost issue. Fourteen scientists from 9 leading research Institutes across Taiwan Strait are invited to give their view and experimental evidence on solving the above problems. These reports highlight the principles of designing delicate microstructure and novel processing routes. The potential of improving machining ability is also demonstrated. However, the most critical cost problem remained to be solved.

In light of the economic surge in China in recent years, the market demands more technology breakthrough. The market can accept value-added products. It further indicates the importance of developing of multi-phase materials with multi-functions. Along with the market growth, the cost problem will be solved soon as the mass production technology becoming available.

We would like to express our gratitude to the contributors of this book for their significant scientific contributions. We would also like to thank Mr. Chin-Lung Hsieh for his effort on the editing of this book. The encouragement and help from Dr. Claus E. Ascheron and his team in Springer Co. are also highly appreciated.

Contents

List of contributors

Bao, Yiwang
China Building Materials Academy, Beijing 100024, China

Guo, Jing-Kun
The State Key Laboratory of High Performance Ceramicsand Superfine Structure, Shanghai Institute of Ceramics, Chinese Academy of Sciences, Shanghai, 200050, China

Hwang, Chii-Shyang
Dept. of Materials Science and Engineering, National Cheng Kung University, Tainan, Taiwan

Huang, Jow-Lay
Dept. of Materials Science and Engineering, National Cheng-Kung University, Tainan, Taiwan

Huang, Yong
Dept. of Material Science and Engineering, Tsinghua University, The State Key Laboratory of New Ceramics & Fine Processing, Beijing, China

Jin, Zhihao
School of Material Science and Engineering, Xi'an Jiaotong University, 710049, Xi'an, China

Pan, You-Bao
Shanghai Institute of Ceramics, Chinese Academy, The State Key Laboratory of High Performance Ceramics and Superfine Microstructure, Shanghai 20050, China

Pan, Wei
Dept. of Materials Science and Engineering, TsingHua University, Beijing 100084, China

Tai, Nyan-Hwa
Dept. of Materials Science and Engineering, National Tsing-Hua University, Hsin-Chu, Taiwan

Tuan, Wei-Hsing
Dept. of Materials Science and Engineering, National Taiwan University, Taipei, Taiwan

Wei, Wen-Cheng
Dept. of Materials Science and Engineering, National Taiwan University, Taipei, Taiwan

Zhang, Dongming
State Key Laboratory of Advanced Technology for Materials Synthesis and Processing, Wuhan University of Technology, Wuhan, 430070, China

Zhang, Lianmeng
State Key Laboratory of Advanced Technology for Materials Synthesis and Processing, Wuhan University of Technology, Wuhan, 430070, China

Zhang, Litong
State Key Laboratory of Solidification Processing, Northwestern Polytechnical University, Xi'an, Shaanxi 710072, China

1 Multi-phase Materials

J.-K. Guo

1.1 Introduction

The trend of ceramic material research has swung back and forth between multi-phase and mono-phase materials in the last 50 years. However, there are many reasons to believe, the ball is coming back to multi-phase materials again.

Fifty years ago, most ceramics were mainly multi-phased materials. For example, the traditional ceramics, such as porcelain, are usually composed of several phases such as glass, quartz, residual feldspar etc. Their microstructure is also complex. It mainly originated from the starting materials used for the ceramics, which are mainly natural minerals. From 1970 onwards the development of advanced ceramics has attracted much attention. The compositions of advanced ceramics are significantly refined, so are their microstructures. The ultimate goal of the microstructural design for the advanced ceramics prepared between 1970 and 1980 is a uniform structure. To achieve this goal, there is usually one phase in the material. However, considering the requirements for structural applications, the strength and toughness that can be achieved by using such mono-phase ceramics are never satisfactory. The need for the development of multi-phase materials has thus been recognized in the last 20 years. Considerable effort has been devoted to various topics on multi-phase ceramics, such as fiber (or whisker) reinforced ceramic matrix composites, particulate-disperse multi-phase ceramics, self-enhanced multi-phase ceramics and composition gradient ceramics.

Significant progress has also been made on strength and toughness enhancement recently. These achievements are based on progress from research areas such as powder synthesis, grain boundary engineering, stress analysis etc. Many hi-tech products are prepared by taking advantage of these achievements. For example, advance in the preparation of carbon fiber has made the space technology in China possible, for which the anti-ablation requirement is critical [1]. Through the design of multi-phase ceramics, the strength of silicon nitride can reach 1000 MPa and maintain its strength up to 1400°C [2,3]. For zirconia-matrix multi-phase ceramics, the strength at room temperature can reach 1500 MPa, and more than 600 MPa when the temperature is above 800°C [4,5]. The strength of mullite multi-phase material can attain 600 MPa and return that to 1000°C [6—8]. Mullite thus has very high potential for ceramic engine applications. The research on multi-phase has proven valuable in terms of performance improvement.

In the last ten years, the study of nano-ceramics has attracted much attention. This development has also given some inspiration to the design of better multi-phase materials. In this book, many examples are given. It has been clearly demonstrated that multi-phase materials can preserve the merits of each composed phase and reduce the weakness of other phases; better performance is thus possible.

1.2 The Basic Concept of Multi-phase Materials

A multi-phase ceramic should contain two phases or more. These phases can be distributed either homogeneously or heterogeneously within the body. The morphology of each phase can also be different. A series of challenges will then be imposed on the design and processing of these materials. These challenges are rooted in the characteristics of the interphase boundary, compatibility, the extent of reactions between phases etc. The strategies needed to solve the problems are outlined below.

1.2.1 Blurring the Boundaries Between Various Phases

Mono-phase materials have many limits due to the characteristics of their microstructures. For example, their grain boundaries always act as the weak parts within the body. For multi-phase material, the grain boundary in the mono-phase material becomes the inter-phase boundary. The boundaries of phases may be blurred due to limited solubility between two nearby phases. Multi-phase materials may not be weak due to the presence of the inter-phase boundary. Furthermore, the performance of the material may not even dominated by the major phase; sometimes the secondary phase may play a more important role. Therefore, we speak of blurring the boundary between different phases. This concept may free the material from the traditional inherent idea, and open up more possibilities.

1.2.2 Applying Nano-technology

For materials with two phases or more, it is most important to tailor their microstructure. It is a challenge to process multi-phase materials where one phase is nano-sized. However, by using nano-material technology, the scale of the microstructure widens from micrometer to nanometer. More alternatives can thus be used to design the microstructure of multi-phase materials and so it is useful to apply nano-fabrication technology to the design of multi-phase materials.

1.2.3 Applying Advanced Fabrication Processes

Several raw materials are usually used for multi-phase materials. The processing routes for advanced materials can also be employed in the processing of multi-phase materials. However, more care is needed. Furthermore, these novel processing routes should aim to "arrange" each phase correctly to achieve a three-dimensional order or randomness.

1.2.4 Marketability and Reliability

The final goal of material research is to apply new materials to the real world. Demand for new materials from the market can push material researchers to move forward. Material scientists should pay more attention to requests from the market, then fulfill the request with new concepts and new materials. To design a new material is the mission of every material scientist. However, it is also the obligation of material researchers to search for economical fabrication routes and make the products as reliable as possible.

1.2.5 Material Design

Two or more functions may be required for one multi-phase material. The amount, composition, morphology etc. of each phase should therefore be carefully monitored. Take the location of one phase as an example: the inter-granular or intra-granular site can make great difference to the performance. If the two materials are layered in two-dimensions, laminated composites can then be designed. The composition in one direction can change from one material to another; a functional gradient material can then be designed.

1.3 The Need for Multi-phase Materials

The development of material marks the progress of human society. Material is the basis as well as the frontier for the development of science and technology. However, to develop a new material is a slow process. In the past, new materials were usually designed by trial and error. However, with the increase of demand for new materials for emerging technologies, mono-phase materials with specific performance no longer fulfill the needs. More new materials with tailorable properties are needed. Our abundant knowledge and experience in using mono-phase materials in the past can help us to design new multi-phase materials. However, we may also have to give up some of the old methodology. Creative thinking is essential to the design of multi-phase materials.

There are several design criteria needed before the start of the design of multi-phase materials. These are:

- The bonding between two phases, should it be strong or weak?
- The reaction between the two phases, is it allowed or not?
- The continuity of the second phase, is it necessary or not?
- For liquid phase processing, is the wetting of the liquid phase on the solid phase critical or not?
- Is it possible to avoid high-temperature processing, such as sintering?

1.4 Research Objectives

Many materials can be chosen for multi-phase material, such as polymer, metal, ceramic. The combination of any two or even three can create many possibilities. Multi-phase materials have applications in areas such as information, energy, environment and biology. The more effort we put into the development of new multi-phase material, the earlier we may be able to see it in use in many applications.

References

1. Guo, J.K., Yan, T.S. (1984) The First China—U.S. Bilateral Seminar on Inorganic Materials Research, ed. by T.S. Yan and J.A. Pask, Science Press, Beijing, China, 281—290
2. Chen, Y., Huang, L.P., Sun, X.W., Jiang, X. (1997) J. Chinese Ceram. Soc., **25**, 183—187
3. Jiang, D.L., She, J.H., Tan, S.H. (1992) J. Amer. Ceram. Soc., **75**, 2586—2589
4. Li, B.S., Huang, X.X., Guo, J.K., Yan, D.S. (1986) J. Inorg. Mater., **1**, 129—134
5. Huang, X.X., Guo, J.K., Gui, L.H., Li, B.S. (1991) Fourth International Symposium on Ceramic Materials and Components for Engines, ed. By Carlsson R., Johansson T., Kahllman L., Goteborg , Sweden, 757—764
6. Huang, X.X., Hong, J.S., Guo, J.K., (1991) Ibid., 795—803
7. Hong, J.S., Huang, X.X., Guo, J.K. (1990) J. Inorg. Mater., **5**, 340—345
8. Hong, J.S., Huang, X.X., Guo J.K. (1996) J. Mater. Sci., **31**, 4847—4852

2 New Concepts in the Design of Tough Ceramics

W.-H. Tuan

2.1 Introduction

The brittle nature of ceramics hinders their applications as structural components. Two alternatives are adopted to solve the problems [1,2]. One alternative refines the processing details to reduce the flaw size in ceramics. The strength is consequently enhanced though the toughness remains low. The other approach targets the toughness improvement through the addition of toughening reinforcement. Ceramic or metallic reinforcement is incorporated into a ceramic matrix. The reinforcement interacts with the pre-existing and/or service-induced cracks to slow down their propagation. The toughness of the brittle matrix is thus improved through such interactions. Though the strength may be sacrificed slightly by adopting this approach, the reinforcement also acts as a stress concentration site. However, the latter approach is attractive, as the reliability of ceramics during the subsequent usage can be improved significantly.

The development of tough ceramic-matrix composites has been thoroughly studied since 1982 [1]. Many ceramic or metallic materials, such as zirconia [3], SiC whisker [2], nickel [4] and silver [5] have been used as toughening reinforcements. The presence of these toughening agents enhances the toughness of ceramics through the generation of various toughening mechanisms. These toughening mechanisms may be active either in a process zone around the major crack or in bridging crack surfaces [6]. The propagation of cracks is hampered due to the effect of these mechanisms, resulting in an increase in the toughness of the ceramics. The presence of a single toughening agent frequently induces more than one toughening mechanism. For example, the addition of zirconia particles can induce transformation toughening, microcrack toughening and crack deflection mechanisms [7]. The metallic inclusions can either bridge the crack surfaces through their plastic deformation [4,5] or deflect the crack [5]. These mechanisms operate simultaneously to a different extent within the brittle matrix. However, the contribution to toughness from each mechanism has yet to be quantified. Several theoretical or experimental attempts have been made recently to differentiate between individual contributions [8,9]. These studies assumed that all the individual contributions to total toughness enhancement could be added together.

This hypothesis is made on the basis that each mechanism interacts with the crack independently. However, it has been demonstrated since that this assumption is not completely true.

The coupling between different toughening mechanisms has received little attention as far. A theoretical prediction [10] suggested that coupling could bring a significant positive contribution to toughness enhancement. This study revealed that the crack-bridging induced by ductile inclusions and transformation toughening induced by zirconia particles can interact. The predicted upper and lower bounds for the toughness ratio, $K_{IC, composite}/K_{IC, matrix}$, can be expressed as [10]
upper bound, for $\chi = \infty$,

$$K_{IC, ductile + transformation} / K_{IC, matrix} = \tag{2.1}$$

$$(K_{IC, ductile} / K_{IC, matrix}) \times (K_{IC, transformation}/K_{IC, matrix})$$

lower bound, for $\chi = 0$,

$$K_{IC, ductile + transformation} / K_{IC, matrix} = \tag{2.2}$$

$$[(K_{IC, ductile} / K_{IC, matrix})^2 + (K_{IC, transformation} / K_{IC, matrix})^2 -1]^{1/2}(1-F)$$

where χ is the coupling parameter between the two mechanisms. This prediction has been supported by an independent experimental study on a mullite matrix composite containing both silicon carbide whiskers and zirconia inclusions [11]. The increase in toughness of composite is indeed higher than the sum of the toughness increase of composites containing a single toughening agent. However, the toughness ratio of the composite is higher than the product of the toughness ratio of the two composites containing only one toughening agent. Several experimental works on various composites have been carried out; but the results contradicted each other [11—16].

In the present study, two toughening agents instead of a single toughening agent are added into an alumina matrix. The interactions between transformation toughening and crack bridging mechanisms are investigated. Zirconia particles were used to generate the transformation toughening mechanism. Nickel and silver inclusions were employed to induce the crack bridging mechanism. The metals are ductile and so deform plastically during crack opening. The plastic deformation contributes to the toughness enhancement [17]. Composites containing only one toughening agent were also prepared for the purpose of comparison.

2.2 Experiments

Detailed procedures for the preparation of the composites containing two toughening agents can be found elsewhere [16]. A brief description is given here. Alumina (TM-DR, Taimei Chem. Co. Ltd., Tokyo, Japan) is mixed with nickel oxide (NiO, Johnson Matthey Co., USA) and zirconia (TZP, ZrO_2 + 3 mol.% Y_2O_3, Hanwha Ceramics Co., Australia) powders by ball milling in ethyl alcohol for 24 hours. The grinding media used were zirconia balls. The slurry of the powder mixtures was dried in a rotary evaporator. The dried lumps were crushed and passed through a plastic sieve. Powder compacts with dimensions of 7 mm× 6 mm× 50 mm were formed by uniaxially pressing at 44 MPa. The green compacts were sintered within a reducing atmosphere, carbon monoxide, at 1600°C for 1 h. The nickel oxide could then be reduced to result in nickel inclusions after sintering [4]. The heating rate and cooling rate were 5°C/min. The Al_2O_3-ZrO_2 and Al_2O_3-Ni specimens, for comparison, were prepared with the same techniques. Some discs of 25.4 mm diameter were prepared for the measurement of elastic modulus with an ultrasonic technique.

The sintered specimens were machined longitudinally with a 325 grit resin-bonded diamond wheel at a cutting depth of 5 μm/pass. The final dimensions of the specimens were 3 mm× 4 mm× 36 mm. The fracture toughness was determined by the single-edge-notched-beam (SENB) technique. The rate of loading was 0.5 mm/minute. The notch was generated by cutting with a diamond saw. The width of the notch was around 0.3 mm. About 8 specimens for each composition were used for the determination of fracture toughness. Phase identification of zirconia on the fracture surface was performed by X-ray diffractometry with CuKα radiation. The relative phase content of zirconia was estimated by using the method proposed by Evans et al. [18]. The final density of the specimens was determined by the Archimedes method. The solubility between the materials used in the present study was low; the relative density of the sintered composites was estimated by using the theoretical density of 3.98 g/cm^3 for Al_2O_3, 6.05 g/cm^3 for TZP and 8.90 g/cm^3 for Ni. Polished surfaces were prepared by grinding and polishing with diamond paste to 6 μm and with silica suspension to 0.05 μm. The metals could vaporize during sintering; the volume fraction of metals in the specimens after sintering was determined by counting the point fraction on the polished surfaces. The polished specimens were then thermally etched at 1500°C for 0.5 h to reveal the grain boundaries of matrix grains. Microstructural characterization was performed using scanning electron microscopy (SEM). The size of Al_2O_3 grains, ZrO_2 and Ni inclusions was determined by using the line intercept technique.

2.3 Results and Discussion

2.3.1 Transformation Toughening Mechanism

The relative density and microstructural characteristics of the Al_2O_3-ZrO_2 composites are shown in Table 2.1. The relative density of the composites sintered at 1600°C is higher than 98%. The presence of ZrO_2 particles prohibits the coarsening of the Al_2O_3 matrix during sintering; the size of Al_2O_3 matrix grains thus decreases with the increase of ZrO_2 content. The size of ZrO_2 particles remains more or less the same with the increase of ZrO_2 content, indicating that ZrO_2 particles are well dispersed within the Al_2O_3 matrix. The coarsening of ZrO_2 particles is thus limited. Since the elastic modulus of ZrO_2 is lower than that of Al_2O_3 (210 GPa vs. 396 GPa), the elastic modulus of the composites decreases with the increase of ZrO_2 content (Fig. 2.1). The elastic modulus, E, of a composite containing both A and B reinforcements can be estimated by using a simple rule of mixtures:

$$E_{composite} = F_{matrix} \times E_{matrix} + F_{A\text{-}inclusion} \times E_{A\text{-}inclusion} + F_{B\text{-}inclusion} \times E_{B\text{-}inclusion} \tag{2.3}$$

where F is the volume fraction. The elastic modulus of Al_2O_3-ZrO_2 composites follows closely to the values predicted by the rule of mixtures, indicating that there is no microcrack existed within the composites. Figure 2.2 shows the toughness of the Al_2O_3-ZrO_2 composites as a function of ZrO_2 content. The extent of phase transformation of zirconia from tetragonal (t) to monoclinic (m) phase on the fracture surface is also shown in the figure. The increase of ZrO_2 content decreases the effective matrix volume around each inclusion, the extent of transformation thus increases with the increase of ZrO_2 content. Both the toughness and the percentage of phase transformation increase with the increase of ZrO_2 content, indicating that the toughness enhancement is mainly contributed by the phase transformation of zirconia. Other toughening mechanisms, such as microcracking toughening and crack deflection, may also be present in the composites; nevertheless, the contribution from these mechanisms to toughness is not significant [7].

Table 2.1. The relative density, size of Al_2O_3 grains and of ZrO_2 inclusions, elastic modulus, toughness and the ratio of the toughness of the composite to that of alumina for Al_2O_3-ZrO_2 composites

Systems	Relative density [%]	Alumina matrix grains [μm]	Zirconia inclusions [μm]	Elastic modulus [GPa]	Toughness [$MPam^{0.5}$]	Toughness ratio
Al_2O_3	99.7	10.2	-	396	3.8±0.1	1.0
Al_2O_3 + 5 vol.% ZrO_2	99.5	2.1	0.27	388	5.0±0.8	1.3
Al_2O_3 + 7.5 vol.% ZrO_2	98.2	2.1	0.34	381	5.2±0.1	1.4
Al_2O_3 + 10 vol.% ZrO_2	99.4	2.0	0.32	381	5.6±0.7	1.5
Al_2O_3 + 12.5 vol.% ZrO_2	99.4	1.7	0.36	375	5.8±0.2	1.5

Table 2.2. The relative density, size of Al_2O_3 grains and of Ni inclusions, elastic modulus, toughness and the toughness ratio for Al_2O_3-Ni composites

Systems	Relative density [%]	Size of Al_2O_3 grains [μm]	Size of metallic inclusions [μm]	Elastic modulus [GPa]	Toughness [$MPam^{0.5}$]	Toughness ratio
Al_2O_3	99.7	10.2	-	396	3.8±0.1	1.0
Al_2O_3 + 4.7vol.% Ni	99.6	6.6	2.4	351	5.2±0.1	1.4
Al_2O_3 + 6.7vol.% Ni	98.7	4.6	2.6	320	5.3±0.6	1.4
Al_2O_3 + 8.7vol.% Ni	99.1	4.4	2.9	285	5.5±0.3	1.5
Al_2O_3 + 10.8vol.% Ni	97.6	4.3	3.7	309	6.0±0.8	1.6
Al_2O_3 + 12.9vol.% Ni	98.1	3.8	4.9	274	6.5±0.4	1.7

2.3.2 Plastic Deformation Mechanism

Table 2.2 collects the values of the properties for the metal-toughened Al_2O_3-Ni composites. The addition of metal inclusions can also prohibit the densification and coarsening of the Al_2O_3 matrix. However, the relative density of the composites is higher than 97.5% as the metal content is lower than 13 vol.%. The size of the metal inclusions increases with the increase of the metal content. Since the elastic modulus of metals is lower than that of alumina (210 GPa for Ni[4] vs. 396 GPa for Al_2O_3), the elastic modulus of the composites decreases with the increase of the metal content. The elastic moduli of the metal-toughened alumina are lower than the values predicted by the rule of mixtures (Fig. 2.1), implying the possible presence of microcracks in the composites.

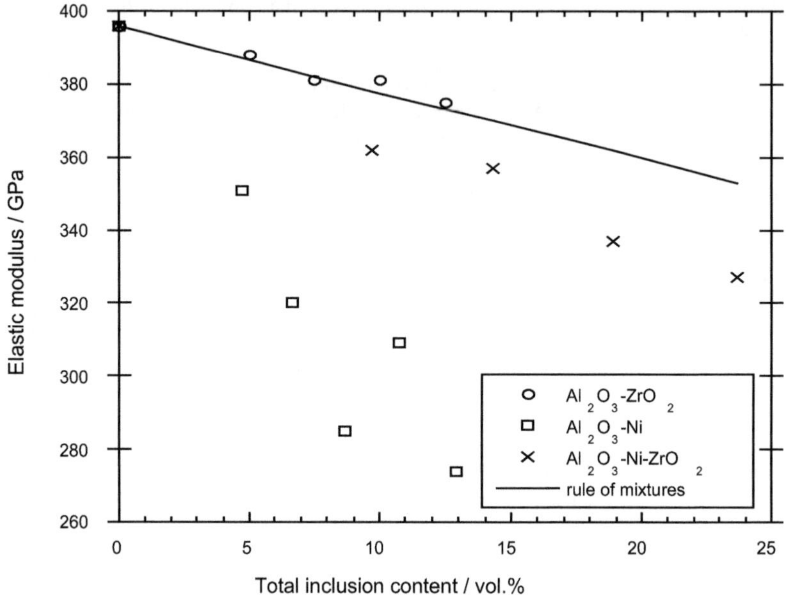

Fig. 2.1. Elastic moduli of composites as a function of total inclusion content. The straight line predicted by the rule of mixtures is shown for comparison

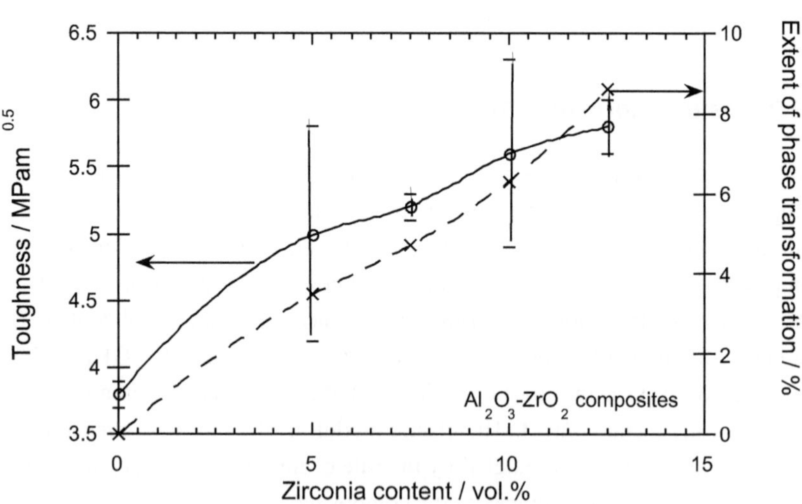

Fig. 2.2. Toughness and the extent of t to m phase transformation on the fracture surface of the Al_2O_3-ZrO_2 composites as a function of ZrO_2 content

Figure 2.3 shows the microstructure of the Al_2O_3-Ni composite. A crack was introduced by Vickers indentation into the polished surface. The crack surfaces were bridged by some stretched metal inclusions, indicating that plastic deformation of metal taken place during the crack opening process. The crack can also propagate along the Al_2O_3/Ni interface.

The addition of nickel inclusions can enhance the toughness of alumina. The increase in toughness, ΔK_{IC}, of the metal-toughened ceramics is proportional to the square root of the product of volume fraction, F, and inclusion size, d, given the toughness enhancement which is mainly contributed by the plastic deformation of metal [17]. The toughness increase, ΔK_{IC}, for the metal-containing composites is shown in Fig. 2.4 as a function of $(F \times d)^{0.5}$. For the Al_2O_3-Ni composites, a linear relationship between ΔK_{IC} and $(F \times d)^{0.5}$, is indeed observed, indicating that the plastic deformation dominates the toughening behavior. Though the crack deflection is also taken place during fracturing, Fig. 2.3, its contribution to toughness enhancement is relatively small.

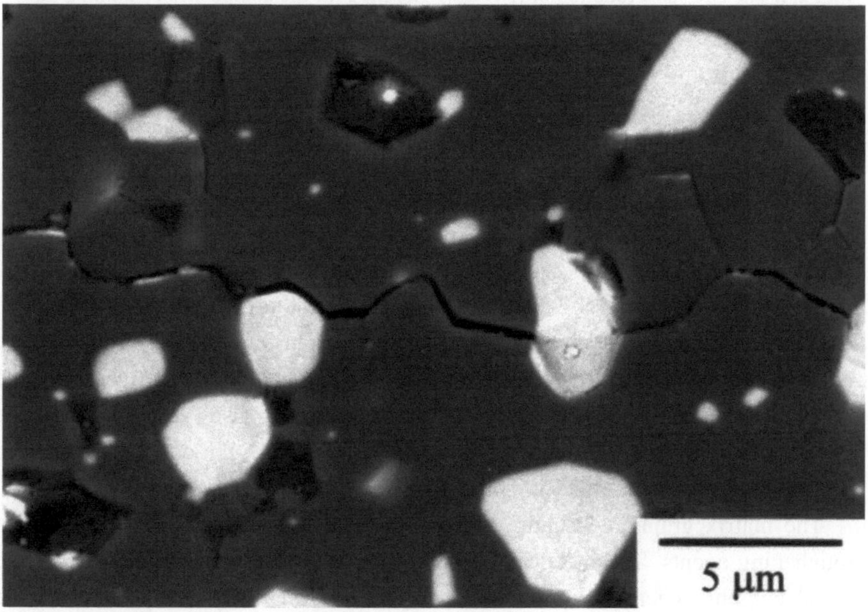

Fig. 2.3. The microstructure of Al_2O_3-12.9%Ni composite (black – Al_2O_3, white – Ni)

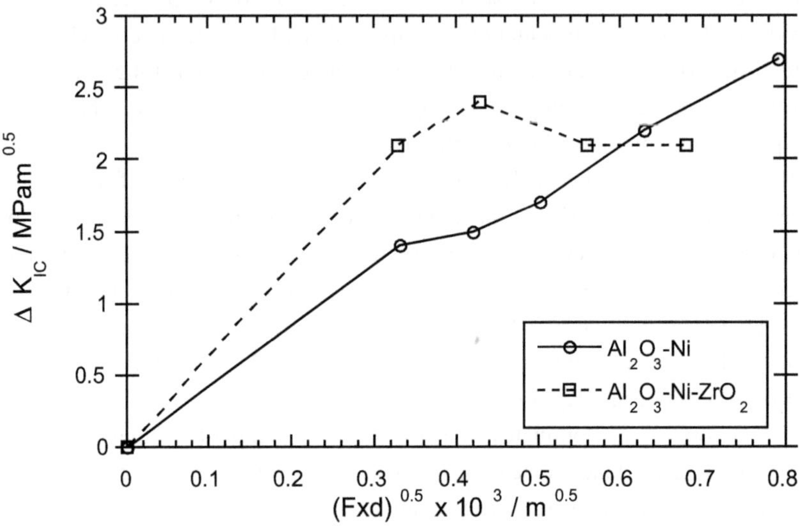

Fig. 2.4. The ΔK_{IC} as a function of $(F \times d)^{0.5}$. The ΔK_{IC} for the composites containing both nickel and zirconia is obtained by subtracting the toughness values of Al_2O_3-ZrO_2 from those of Al_2O_3-Ni-ZrO_2 composites in Fig. 2.8 in terms of the same extent of phase transformation

2.3.3 Interactions Between Transformation and Toughening Plastic Deformation

Table 2.3 shows the properties of the composites containing both zirconia and metal inclusions. Figure 2.5 shows the microstructure of the Al_2O_3-Ni-ZrO_2 composite. The nickel and zirconia inclusions are distributed uniformly within the alumina matrix. The crack surfaces induced by indentation are bridged by some metal inclusions, indicating that crack bridging is also active within the composites. The Al_2O_3/Ni interface is relatively weak [4]; the crack can thus also propagate along the interface of some inclusions.

The matrix grains are significantly reduced in size after the addition of two toughening agents. The metal inclusions increase in size with the increase of the metal content, whereas the zirconia particles are relatively independent of the zirconia content. The elastic moduli of the composites decrease with the increase of the total inclusion content; furthermore, the experimental values are slightly lower than the values predicted by the rule of mixtures. From Fig. 2.1, the addition of ZrO_2 into Al_2O_3-Ni composites can nevertheless increase their stiffness. The coefficient of the thermal expansion of ZrO_2 locates between that of Al_2O_3 and Ni; the addition ZrO_2 thus reduces the thermal residual stresses within the matrix.

Table 2.3. The relative density, size of Al_2O_3 grains and of Ni and ZrO_2 inclusions, elastic modulus, toughness and the toughness ratio for Al_2O_3-Ni-ZrO_2 composites

Systems	Relative density [%]	Size of Al_2O_3 grains [μm]	Size of ZrO_2 particles [μm]	Size of metallic inclusions [μm]	Elastic modulus [GPa]	Toughness [$MPam^{0.5}$]	Toughness ratio
Al_2O_3	99.7	10.2	-	-	396	3.8±0.1	1.0
Al_2O_3 + 4.7vol.% Ni + 5vol.% ZrO_2	99.7	1.9	0.35	2.3	362	7.9±0.1	2.1
Al_2O_3 + 6.7vol.% Ni + 7.5vol.% ZrO_2	99.2	1.8	0.29	2.8	357	9.1±0.5	2.4
Al_2O_3 + 8.8vol.% Ni + 10vol.% ZrO_2	98.6	1.6	0.32	3.6	337	9.5±0.1	2.5
Al_2O_3 + 11vol.% Ni + 12.5vol.% ZrO_2	98.8	1.5	0.33	4.3	327	10.0±0.7	2.6

Smaller thermal stresses result in fewer microcracks; which raises the elastic moduli of the composites containing both metals and zirconia.

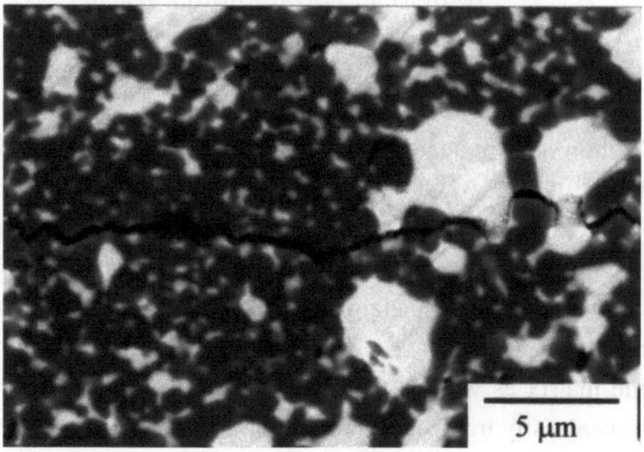

Fig. 2.5. The microstructure of Al_2O_3-11.0% Ni-12.5% ZrO_2 composite (black – Al_2O_3, white – Ni, gray – ZrO_2)

Figure 2.6 shows the toughness ratio of the Al_2O_3-Ni-ZrO_2 composites as a function of the total inclusion content. The upper bound and lower bound for the toughness ratio as calculated by (2.1) and (2.2) are also shown in the figure. The experimental values are even higher than the product of the toughness ratio of the composite containing only nickel and only zirconia.

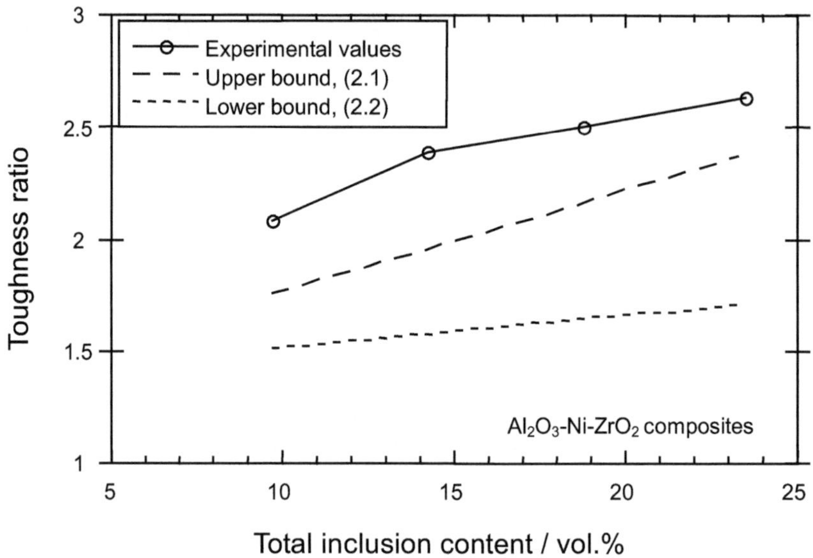

Fig. 2.6. The toughness ratio of Al_2O_3-Ni-ZrO_2 composites as a function of total inclusion

Figure 2.7 shows the extent of zirconia t to m phase transformation on the fracture surfaces of the Al_2O_3-Ni-ZrO_2 composites as a function of zirconia content. The percentage of phase transformation for Al_2O_3-ZrO_2 composites is also shown to provide a basis for comparison. More ZrO_2 particles transform from t to m phase due to the presence of metal inclusions, indicating that the transformation toughening mechanism is enhanced. The addition of metal inclusions reduces the stiffness of the matrix as well as the constraint on the zirconia particles, so that these particles can easily transform from t to m. Figure 2.8 shows the toughness of the ZrO_2-containing composites as a function of the percentage of phase transformation. The toughness of the composites containing both metal and ZrO_2 is higher than that of the composite containing only ZrO_2 in terms of the same amount of ZrO_2, indicating that other mechanisms are also active in the composites. From microstructural observation (Fig. 2.5) the plastic deformation of Ni has also taken place during the fracturing process.

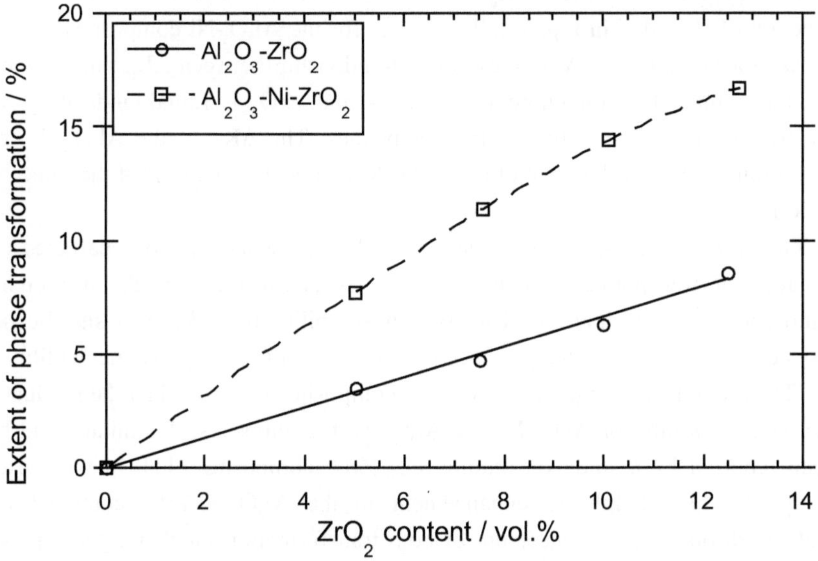

Fig. 2.7. The extent of ZrO_2 t to m phase transformation on fracture surface as a function of ZrO_2 content

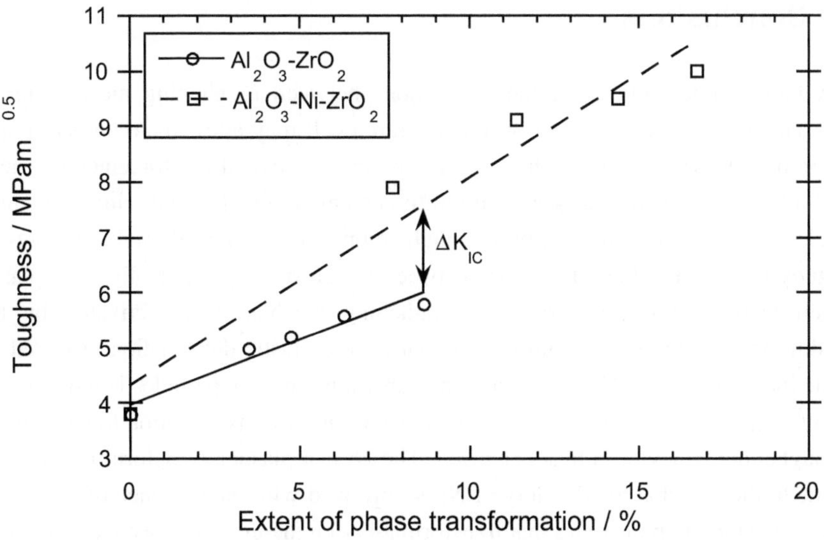

Fig. 2.8. The toughness of ZrO_2-containing composites as a function of the extent of phase transformation

The toughness increase, ΔK_{IC}, for Al_2O_3-Ni-ZrO_2 composites is shown as a function of $(F \times d)^{0.5}$ in Fig. 2.4. The values for the Al_2O_3-Ni composites are also shown for comparison. A roughly linear relationship between ΔK_{IC} and $(F \times d)^{0.5}$ for the composites containing both metals and ZrO_2 is found, indicating that plastic deformation is active in the composites. The ΔK_{IC} of the Al_2O_3-Ni-ZrO_2 composites is higher than that of the Al_2O_3-Ni composites in terms of the same $(F \times d)^{0.5}$.

Zirconia is a good oxygen conductor. The presence of ZrO_2 particles can enhance the transportation of the dissolved oxygen in nickel to the surrounding atmosphere during sintering. The oxygen solubility in nickel can significantly reduce its ductility [20]; the presence of zirconia can thus improve the ductility of Ni. The toughness ratio of Al_2O_3-Ni-ZrO_2 composites is higher than the product of the toughness ratio of Al_2O_3-Ni and Al_2O_3-ZrO_2 composites. A similar trend has been reported by a study on mullite-SiC_w-ZrO_2 composites [11]. In the present study, the high toughness enhancement in the Al_2O_3-Ni-ZrO_2 composites is realized through the enhancement of both transformation toughening and plastic deformation. It suggests that there is a need for more theoretical and experimental studies on composites containing multiple toughening mechanisms.

4 Conclusions

A better understanding of the interactions between toughening mechanisms is essential to the development of tough ceramics. It may allow us to prepare tough composites with the addition of small amounts of several reinforcements instead of a large amount of one particular reinforcement. While the total reinforcement is relatively low the strength of the tough composite may be still high. The present study investigates the interactions between transformation toughening and plastic deformation. The transformation toughening mechanism is activated by the addition of ZrO_2 particles into Al_2O_3 matrix; the plastic deformation provided by the incorporation of Ni inclusions. The constraint on ZrO_2 particles is reduced due to the addition of ductile nickel. The contribution of transformation toughening to toughening behavior is thus enhanced. The Ni can plastically deform more easily due to the fact that the ductility of Ni is improved with the presence of ZrO_2. The present study demonstrates that transformation toughening and plastic deformation do interact with each other to some extent. The interactions between two toughening mechanisms thus play an important role on the toughening behavior. Such interactions depend strongly on the materials characteristics and microstructure features.

Acknowledgement

The National Science Council, R.O.C., supported the present study through the contract number of NSC89-2216-E002-049. The technical help from Dr. Rong-Zhi Chen is highly appreciated.

References

1. Evans, A.G. (1990) J. Am. Ceram. Soc., **73**, 187
2. Brook, R.J. (1989) J. Euro. Ceram. Soc., **5**, 75
3. Claussen, N. (1984) Advances in Ceramics, 1984, Vol. **12**, pp. 325, ed. by Claussen, N., Ruhle, M., Heuer, A.H., Am. Ceram. Soc., Columbus, USA
4. Tuan, W.H., Brook, R.J. (1990) J. Euro. Ceram. Soc., **6**, 31
5. Chou, W.B., Tuan, W.H. (1995) J. Euro. Ceram. Soc., **15**, 291
6. Ruehle, M., Evans, A.G. (1989) Progress in Mater. Sci., **33**, 85
7. Hannink, R.H.J., Kelly, P.M., Muddle, B. C. (2000) J. Am. Ceram. Soc., **83**, 461
8. Zhai, H., Huang, Y., Wang, C., Wu, X. (2000) J. Am. Ceram. Soc., **83**, 2006
9. Shin, Y-S, Rhee, Y-W., Kang, S-J L. (1999) J. Am. Ceram. Soc., **82**, 1229
10. Amazigo, J.C., Budiansky, B. (1988) J. Mech. Phys. Solids, **36**, 581
11. Becher, P.F., Tiegs, T.N. (1987) J. Am. Ceram. Soc., **70**, 651
12. Claussen, N., Petzow, G. (1986) J. de Phys., Colloque C1, supplement au n2, Tome **47**, c1— 6693
13. Jang, H.M., Moon, J.H., Jang,, C.W. (1992) J. Am. Ceram. Soc., **75**, 3369
14. Hong, J.S., Huang, X.X., Guo, J.K. (1996) J. Mater. Sci., **31**, 4847
15. Ruh, R., Mazdiyasni, K.S., Mendiratta, M.G. (1988) J. Am. Ceram. Soc., **71**, 503
16. Chen, R.Z., Tuan, W.H. (2000) J. Euro. Ceram. Soc., **20**, 1901
17. Ashby, M.F., Blunt, F.J., Bannister, M. (1989) Acta Metall., **37**, 1847
18. Evans, P.A., Stevens, R., Binner, J.P. (1984) Brit. Ceram. Trans. J., **84**, 39
19. Davidge, R.W., Green, T.J. (1968) J. Mater. Sci., **3**, 629
20. Hoerz, G., Speck, H., Hehn, H., Formm E., Jehn, H. (1983) in Physics Data, Gases and Carbon in Metals, Max-Planck Inst., Stuttgart, Germany

3 Interface Design of Tough Ceramics : Si_3N_4/BN Fibrous Monolithic Ceramic

Y. Huang, L. Zou, S. Li, C. Wang, C. Li, Q. Zan

3.1 Introduction

The applications of silicon nitride are restricted mainly by its brittleness. This is why so much attention paid to the toughness enhancement of ceramics. In the beginning of 1990's, Clegg [1] proposed a process to fabricate a laminated SiC/C composite, which possessed much higher fracture toughness and work of fracture. This provided a new approach to ceramic toughening. After the idea, Coblenze [2], Baskaran [3—5] succeeded in synthesizing in-situ fibrous mônolithic ceramics. Since 1993, this subject has also been worked on by the present authors. The idea of biomimetic structural design was also incorporated into the basic concept [6].

In applying this idea, the bulk ceramic materials consisted of polycrystalline Si_3N_4 ceramic fibers, and interphases were successfully synthesized in-situ by simulating the structural characteristics of wood and bamboo. As the material prepared in this way is loaded, cracks will deflect and propagate in the weak interphases, subsequently kinking out of one interphase and deflecting into another one. This process consumes a large amount of fracture energy, and reduces the dependence of mechanical properties on its original crack, i.e. reduces the sensitivity of the crack. The properties of Si_3N_4/BN fibrous monolithic ceramics depend strongly on their interfacial bonding strength. It has been demonstrated both theoretically and experimentally that there must be an appropriate interfacial bonding strength for the ceramic material to achieve higher fracture toughness and work of fracture, as well as a higher strength. Usually, interfacial toughness, i.e. interfacial strain energy release rate or interfacial fracture resistance, is needed to characterize the interfacial bonding strength. Unfortunately, the interfacial toughness data of ceramic materials were seldom available, and so far there is no reliable method to measure and characterize this parameter. The interface bonding state can only be evaluated by examining the final properties of the materials. It is thus impossible to start from interfacial design to obtain the materials with expected properties. Due to the lack of these data, material design and numerical simulation cannot be performed. Furthermore, because the data cited from literatures are not reliable, they will result in a significant deviation between the calculated values and real values.

Starting from the design of interfacial composition, this study presents a method to measure and characterize interfacial toughness of Si_3N_4/BN composite. Using this method, the interfacial toughness of the BN+Al_2O_3 interphase system with different Al_2O_3 volume percentages was measured, then the best interfacial composition was determined. The corresponding Si_3N_4/BN fibrous monolithic ceramics were fabricated based on this composition. The properties of high-temperature creep and thermal shock resistance were also investigated.

3.2 Measurement of Interfacial Toughness

Based on the three-point bending fracture model proposed by Phillips et al [7] and bimaterial interface model by Charalambides et al. [8], we proposed a method by which it is easy to characterize nterfacial toughness [9]. The Si_3N_4/BN/Si_3N_4 sandwich specimen with a single interlayer was prepared. A crack was introduced to one side of the specimen only. The crack was connected to an interphase located at the middle of Si_3N_4 matrix. The specimens were loaded by four-point bending with an outer span of 40 mm and an inner span of 20 mm (Fig. 3.1). The loading process was controlled by displacement. The corresponding load-displacement curves were obtained, and finally the interfacial toughness was calculated.

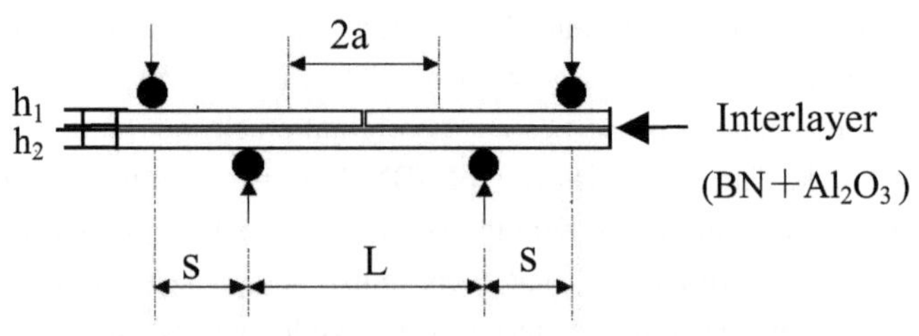

Fig. 3.1. The schematic for the interphase toughness measurement

The results corresponding to specimens with different interfacial composition of BN+Al_2O_3 interphase are illustrated in Fig. 2 and Table 1. It shows that interfacial toughness increases with increase of volume percentages of Al_2O_3, which indicates that the interfacial bonding is gradually enhanced with the increase of Al_2O_3 modifier. When the interphase is completely composed of pure Al_2O_3, the crack cannot deflect in the interphase. It initiates a crack in Si_3N_4 matrix and results in a brittle fracture (Fig. 3.2d). From Table 3.1, it also shows that fracture exhibits a maximum as the added Al_2O_3 is 36 vol.%. To improve the toughness of Si_3N_4/BN composite without sacrificing its strength, BN+36vol.% Al_2O_3 was chosen as the interfacial composition .

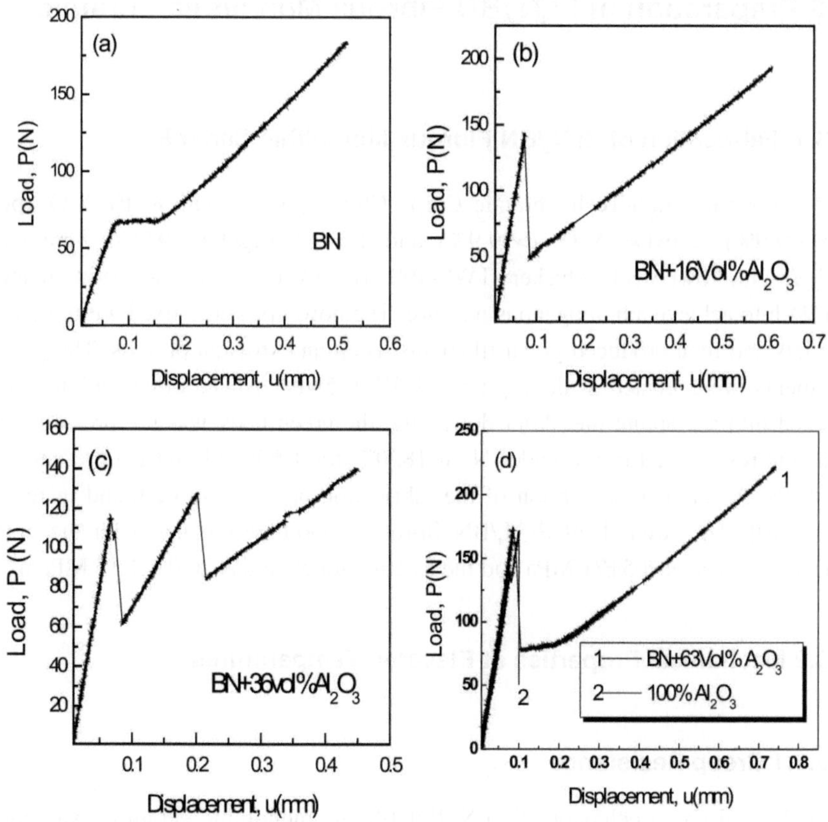

Fig. 3.2. The load-displacement curves of Si_3N_4/BN/Si_3N_4 sandwich specimens with different amounts of Al_2O_3 modifier

Table 3.1. The values of interfacial toughness and work of fracture (WOF) for the $Si_3N_4/BN/Si_3N_4$ specimens with different interphase compositions

Parameters	Amount of Al_2O_3 [vol.%]				
	0	16%	36%	63%	100%
G_i [J/m²]	37.16	53.95	71.86	119.21	-
WOF[J/m²]	2880	3302	5255	4137	268

3.3 Preparation of Si_3N_4/BN Fibrous Monolithic Ceramic

3.3.1. Fabrication of Si_3N_4/BN Fibrous Monolithic Ceramic

Si_3N_4 (Founder High-Tech Ceramic Corp., China) powder with 8wt% Y_2O_3 (purity>99.9%), 2.5wt% Al_2O_3 (>99.9%) and 1.5wt% MgO (>99.9%) were ball milled with 20wt% SiC whisker (TWS-400, Tokai Carbon Co., Japan) in ethanol for 24 h to achieve a homogenous mixture. The powders were mixed with organic binders and then produced green filaments using an extrusion process. The green filaments were coated with a slurry of BN+75wt% Al_2O_3, dried and parallel packed into a graphite die. After dewaxing, the green body was hot pressed in a graphite resistance furnace under N_2 at 1820°C for 1.5 h and under a pressure of 22 MPa. A detailed description of the fabrication process can be found in reference [10]. The strength of Si_3N_4/BN fibrous monolithic ceramic with the optimized interface is 705±71 MPa and the fracture toughness is 20.01±1.17 $MPam^{1/2}$.

3.3.2 Mechanical Properties at Elevated Temperatures

3.3.2.1 Creep Resistance

The flexural creep behavior of Si_3N_4/BN fibrous monolithic ceramics was conducted by using a creep tester in a temperature ranges from 1000°C to 1200°C

Table 3.2. The rupture times of the composites crept at different temperatures under various stresses

Temperature [°C]	Stress [MPa]	Rupture times [Hours]
1000	250	>300
	350	>300
	400	>300
	500	50
	600	2
1100	250	>300
	350	>300
	400	>300
	500	16
	600	1.2
1200	250	>300
	350	210
	400	40
	500	4
	600	0.7

Center-point deflections during the creep test were measured directly using a LVDT. The heating rate was 500°C/h. The specimens were held at the testing temperature for 15 minutes before the load was applied. Table 3.2 shows the rupture times of the composites at different temperatures under a variety of stresses. The typical creep curves at 1200°C under different stresses are shown in Fig. 3.3. From the experimental results, it can be seen that the materials exhibit excellent creep resistance. The reasons are as follows: (1) The BN interlayer in Si_3N_4/BN fibrous monolithic ceramics has the capacity to absorb the glass from the Si_3N_4 grain boundary. (2) The structure of the BN interlayer is loose and easily absorbs the deformation during creep. (3) SiC whiskers were added, which produce pulling-out effects to improve the creep resistance.

3.3.2.2 Thermal Shock Behavior

The furnace was heated up to the testing temperature, the specimens were then put into it and that temperature was maintained for 20 min. Then the samples were dropped into water. The retained strength of the materials was measured at room temperature. The retained strength as a function of the temperature difference is shown in Fig. 3.4. The results reveal that the bending strength of the Si_3N_4/BN fibrous monolithic ceramic degraded abruptly above 700°C. The thermal shock critical temperature ΔTc is thus 700°C, which is 100°C higher than that of Si_3N_4 monolithic ceramics [11].

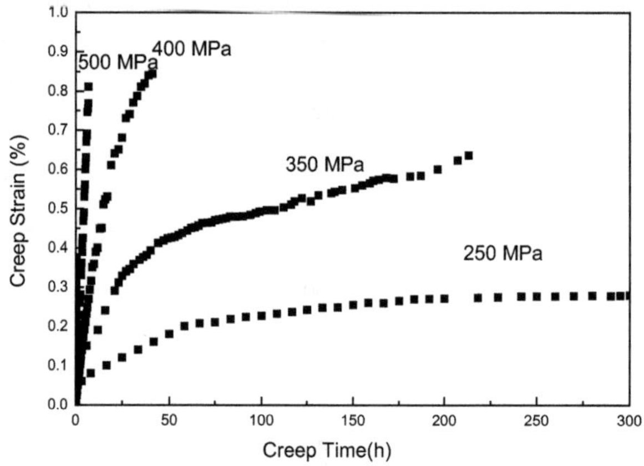

Fig. 3.3. The typical creep curves of Si_3N_4/BN fibrous monolithic ceramics under different stresses

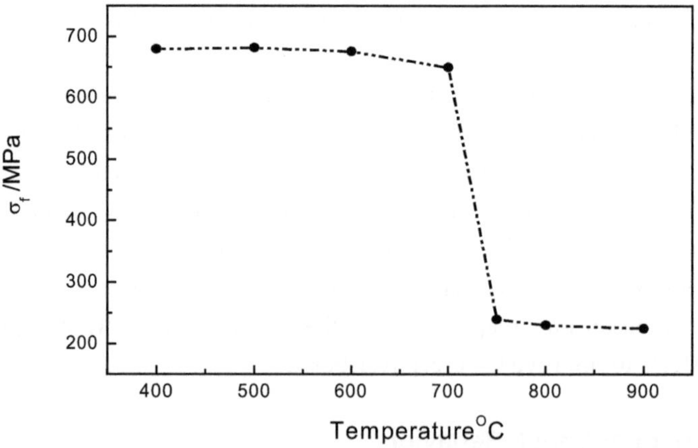

Fig. 3.4. The thermal shock resistance of Si_3N_4/BN fibrous monolithic ceramics

It is known that a thermal shock weakens the strength significantly due to the formation of surface cracks. In order to prevent crack formation, it is necessary to have high strength and thermal conductivity but low Poisson ratio, thermal expansion coefficient and elastic modulus. In addition, the ability to inhibit the crack propagation is also important, which can be attained by increasing the fracture energy, elastic modulus, and Poisson ratio. It is noted that strength, elastic modulus, and Poisson ratio have an adverse effect on crack initiation and propagation. However, the improvement of fracture energy is beneficial to thermal shock resistance [12]. It is known that BN is a weak interlayer. When a crack extends to the interlayer, the crack deflection as well as delamination and crack branching occur at the Si_3N_4 boundaries. These can all absorb energy and improve the thermal shock resistance. The sintering additives, Al_2O_3 and Y_2O_3, added to the Si_3N_4/BN fibrous monolithic can form a sialon boundary phase, which is helpful to the thermal shock resistance [13]. Moreover, during the sintering process, both Al and Y are easy to diffuse into the BN interlayer, so the Si_3N_4 boundary is purified [14,15] and the properties at elevated temperatures are improved. The SiC whiskers have due to great effect on the thermal shock resistance for their high thermal conductivity [16]. As the whiskers are added, the thermal conductivity of the Si_3N_4/BN fibrous monolithic ceramics can be increased. The temperature gradient in the specimen becomes small, which leads to a decrease of the thermal stress. So the retained strength of the specimen after quenching is thus higher than that of the matrix material. Moreover, the whiskers can produce crack deflection, bridging and pull-out. The fracture energy and the thermal shock resistance of the composites are increased [17].

3.4 Conclusions

Several conclusions can be drawn from the present study.

1. Interphase toughness plays an important role in the performance of ceramic composites. A method was developed to characterize the interfacial toughness of Si_3N_4/BN. The method was proved to be effective and viable.
2. The composition of BN+36vol.% Al_2O_3 for interphase can result in the highest interfacial toughness. The Si_3N_4/BN fibrous monolithic ceramic materials with this interphase exhibit the strength and toughness of 705±71 MPa and 20.0±1.2 $MPam^{1/2}$, respectively.
3. The Si_3N_4/BN fibrous monolithic ceramic exhibits excellent properties at elevated temperatures. The improvement can be related to the purified Si_3N_4 grain boundary, the presence of a soft BN interlayer and the addition of SiC whiskers.
4. The thermal shock critical temperature ΔTc of fibrous monolithic ceramic is 700°C, which is 100°C higher than that of Si_3N_4 monolithic ceramics. The enhancement is contributed by the higher work of fracture.

Acknowledgement

This work was supported by National Natural Science Foundation of China (NSF) under grant no. 59632090.

References

1. Clegg, W.J., Kendall, K., Alford, N.M. (1990) Nature, **347**, 445
2. Coblenz, W.S. (1988) U.S. Pat. No. 4772524, September 20
3. Baskaran, S., Nunn, S., Popovic, D., Halloran, J.H. (1993) J. Am. Ceram. Soc., **76**, 2209
4. Baskaran, S., Halloran, J.W. (1993) J. Am. Ceram. Soc., **76**, 2217
5. Baskaran, S., Halloran, J.W. (1994) J. Am. Ceram. Soc., **77**, 1249
6. Yong, H., Wang, C. (2000) Proceedings of first Sino-Taiwan Conference on Ceramic and Metal Matrix Composite (in Chinese), Shanghai, 68
7. Phillips, A.J., Clegg, W.J., Clyne, T.W. (1993) Acta Metall. Mater., **41**, 805
8. Charalambidees, P.G., Lund, J., Evans, A.G., McMeeking, R.M. (1989) J. Appl. Mech., **56**, 77
9. Zou, L., Huang, Y., Wang, C. (in press) Journal of the Chinese Ceramic Society (in Chinese)
10. Guo, H., Huang, Y., Wang, C. (1999) J. Mater. Sci., **34**, 2455
11. Dutta, S. (1991) Key Engineering Materials, **56—57**, 99
12. Zhang, B., Guo, J., Zhu, P., Huang, X. (1995) Ceramic Bulletin (in Chinese), **3**, 35
13. Ma, N., Lan, J., Shi, Z. (1997) Ceramic Research (in Chinese), **12**, 7
14. Trice, R.W., Halloran, J.W. (2000) J. Am. Ceram. Soc., **83**, 311
15. Huang, Y., Guo, H., Xie, Z.P. (1998) J. Mater. Sci. Lett., **17**, 569
16. Jiang, D. (1995) Journal of Inorganic Materials (in Chinese), **10**, 151
17. Becher, P.F., Tiegs, T.N. (1988) Adv. Ceramic. Mater, **3**, 148

4 Microstructural Design : Si/SiC Composites Derived from Wood

G. Qiao, Z. Jin

4.1 Introduction

Wood is an important renewable resource, which has a unique microstructure. This structure results in very good mechanical properties [1]. Although it has a considerable amount of porosity, wood exhibits a remarkable combination of high strength, stiffness and toughness due to its cellular structure [2,3]. Advanced high-temperature, cellular materials can be prepared by using biological preforms as structural templates for ceramic composites [4].

Charcoal is old material made from natural wood or wood fiber and has been used as fuel for many years. Woodceramics are porous carbon materials with excellent environmental properties. These ceramics are made from natural wood or wood fiber. They have many impressive characteristics, such as high electric conductivity and self-lubricity, and can be used as friction material and electromagnetic shield material [5]. However, woodceramics can not be applied at higher temperatures (>600°C) as structural materials, as other "ceramics" can, because of their lower strength and poor oxidation resistance [6]. So the applications of woodceramics are very limited.

Silicon carbide is an important structural ceramic material with excellent high-temperature properties, including high strength, oxidation resistance and good thermal shock resistance. A type of SiC product, named "Reaction Sintered SiC" is made from the powder mixture of carbon and silicon by a chemical reaction:

$$Si+C = SiC \tag{4.1}$$

SiC may also be fabricated by the high-temperature reaction between porous woodceramics and liquid silicon.

A new technique was developed to prepare Si/SiC ceramic composites or porous SiC with natural wood as the main starting material [1,4,6—21]. The processing, properties and microstructure of the composites are introduced in the present study.

4.2 Experimental Work

4.2.1 Starting Materials

Pine, birch, bamboo and MDF (medium density fiberboard) were used as carbon sources. The wood was cut into square pieces and dried at 120°C for more than 24 h. Industry silicon powder with a diameter of 1–5 mm was used as a silicon source.

4.2.2 Fabrication Process

The four types of wood were first pyrolyzed at 900°C for 4 h with N_2 protection to prepare charcoal. The heating rate should be slower than 5°C min^{-1} to avoid distortion or cracking of charcoal pieces. The charcoal pieces were packed in silicon powder and heated together at 1600°C for 1 h in a vacuum furnace. Finally S/SiC ceramics were prepared through the reaction between charcoal and liquid silicon. The fabrication process is shown in Fig. 4.1.

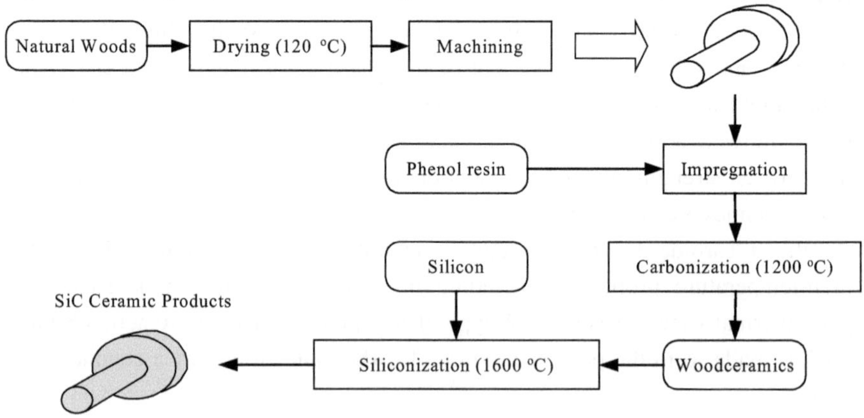

Fig. 4.1. Process flowchart for the preparation of Si/SiC ceramics from wood

4.2.3 Testing

Prepared Si/SiC was cut into specimen bars with size of 4 mm × 5 mm × 40 mm. The cutting direction was selected carefully so that the long edge was parallel with the fiber growth direction of natural wood. Strength was tested by the three-point bending method with a 30 mm span. Fracture toughness was tested with the SENB method. A notch with 2 mm depth and 0.28 mm width was cut in the middle of each SENB specimen. The microstructure of specimens was observed by optical microscope and SEM. X-ray diffraction (XRD) was used to determine the phase composition.

4.3 Results and Discussion

4.3.1 XRD Analysis and Phase Composition

Fig. 4.2 is an XRD pattern of an Si/SiC ceramic specimen fabricated from lacebark pine. XRD patterns of SiC product from the other three types of wood are very similar to Fig. 4.2. It can be seen that ceramics prepared by this new technique are a multi-phase material, which consists of β-SiC and free silicon. The density and porosity of charcoal and Si/SiC specimens were determined through the Archimedes method. A density calculation method was used to determine the phase composition [4]. The calculated results are listed in Table 4.1.

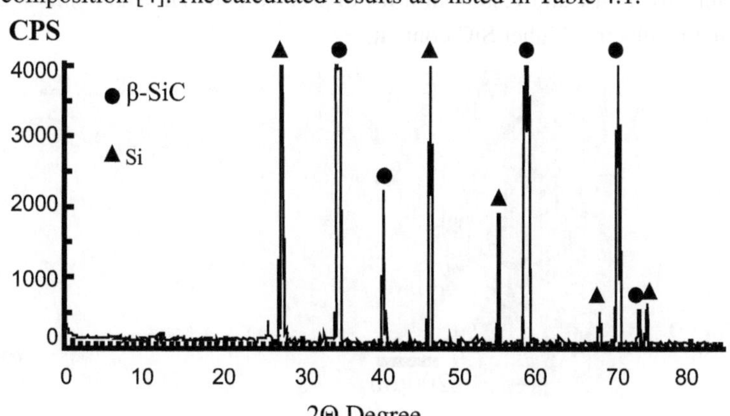

Fig. 4.2. XRD pattern of the final product

From Table 4.1, it can be seen that Si/SiC ceramics fabricated by this method contain a large quantity of free-silicon. In the SiC ceramics made from pine, MDF and birch wood, the absence of free carbon indicates that carbon and silicon react thoroughly under these experimental conditions. On the other hand, more than 20 vol.% of free carbon was found in SiC ceramics made from bamboo. This may be because there are many bulky fiber-pipes in bamboo which turned into carbon

Table 4.1. Phase composition of Si/SiC ceramics

Original wood type	Bamboo	Pine	MDF	Birch
Density [g cm^{-3}]	2.44	2.62	2.70	2.74
SiC content [vol.%]	26.3	31.9	42.8	45.9
Free silicon [vol.%]	52.6	68.1	57.2	54.0
Free carbon [vol.%]	20.6	~0	~0	~0
Pore [vol.%]	0.51	0.02	0.01	0.09

pipes with thick walls after pyrolysis. These thick-walled carbon pipes cannot react with silicon thoroughly.

4.3.2 Microstructure

Figure 4.3 shows the optical micrograph of Si/SiC ceramics derived from pine and MDF. The gray blocks are SiC grains from the Si-C reaction. The white blocks are free silicon not reacted. It can be seen that ceramics made from pine have a fiber-type anisotropic structure similar to that of natural wood but those from MDF have an isotropic structure. The color difference between the left and right sides in Figure 4.3a reflects the growth ring in wood. The left side is a loose layer while the plant grows fast and results in a lot of free silicon. The right side is a dense layer and results in a higher SiC content.

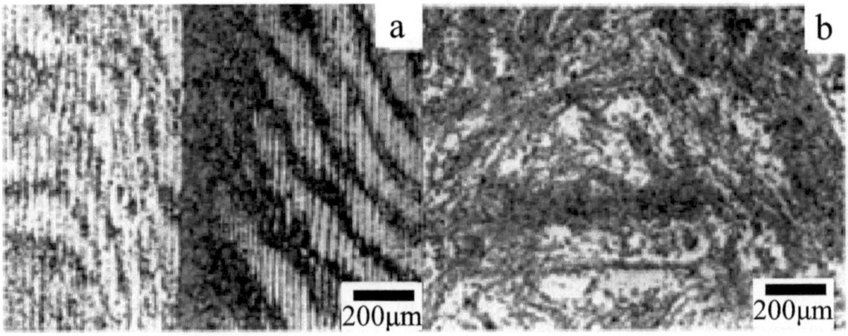

Fig. 4.3. Optical micrographs of Si/SiC ceramics derived from (a) pine (parallel to the growing direction) and (b) MDF

Figure 4.4 shows the optical micrograph of Si/SiC ceramics made from birch. The small gray blocks are β-SiC grains from the Si-C reaction. The small white blocks are the free silicon in cells, and the large white blocks are the free Si in pipes. A special anisotropic structure can be seen clearly, which is very similar to that of the original wood. In the pipe-scale (~100 μm), good duplication of microstructure can be observed among the original wood, woodceramic and final SiC ceramic. But in the cell-scale (~10 μm), no clear similarity of microstructure can be observed. Cell structure in the original wood and woodceramic was destroyed during the Si/C reaction at high temperature, and only typical granular SiC and free Si were observed.

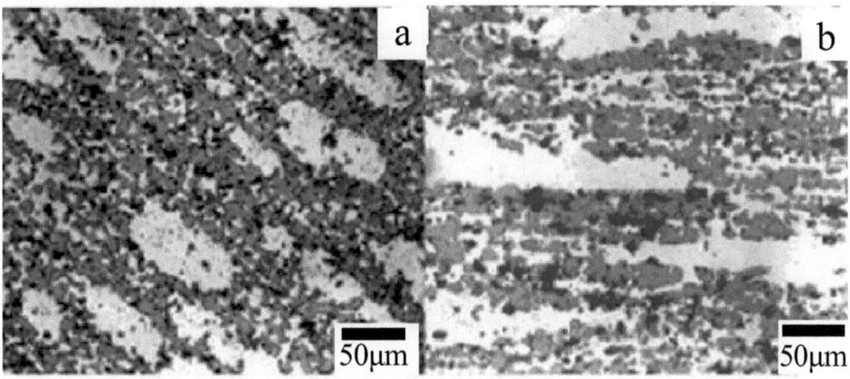

Fig. 4.4. Optical micrographs of Si/SiC ceramics derived from birch (a) perpendicular and (b) parallel to the growing direction

4.3.3 Mechanical Properties

Figure 4.5 shows the bending strength of Si/SiC ceramics derived from the four types of wood as a function of SiC content. Both the strength data and SiC content varied in a wide scope because the uniformity of the original wood pieces was poor. This phenomenon was observed widely and may be the regular rule for biomaterials. Although the strength data give a large scatter, it can be seen clearly that the strength increases with the increase of SiC content. But the MDF is an exception to this trend. The ceramics derived from MDF have higher SiC content and give lower strength, compared with ceramics made from pine. MDF is a manmade material derived from sawdust and has lower macroscopic anisotropy. On the other hand, pine, bamboo and birch are all natural woods with a large quantity of long fibers and pipes, so they show distinct anisotropy. Therefore, there may be remarkable differences between the microstructure of Si/SiC ceramics made from natural wood and MDF. This will be discussed further in the following section. Figure 4.5 also shows the fracture toughness of Si/SiC ceramics derived from the four types of wood as a function of SiC content. The toughness also increases with the increase of SiC content, except the Si/SiC ceramics derived from MDF.

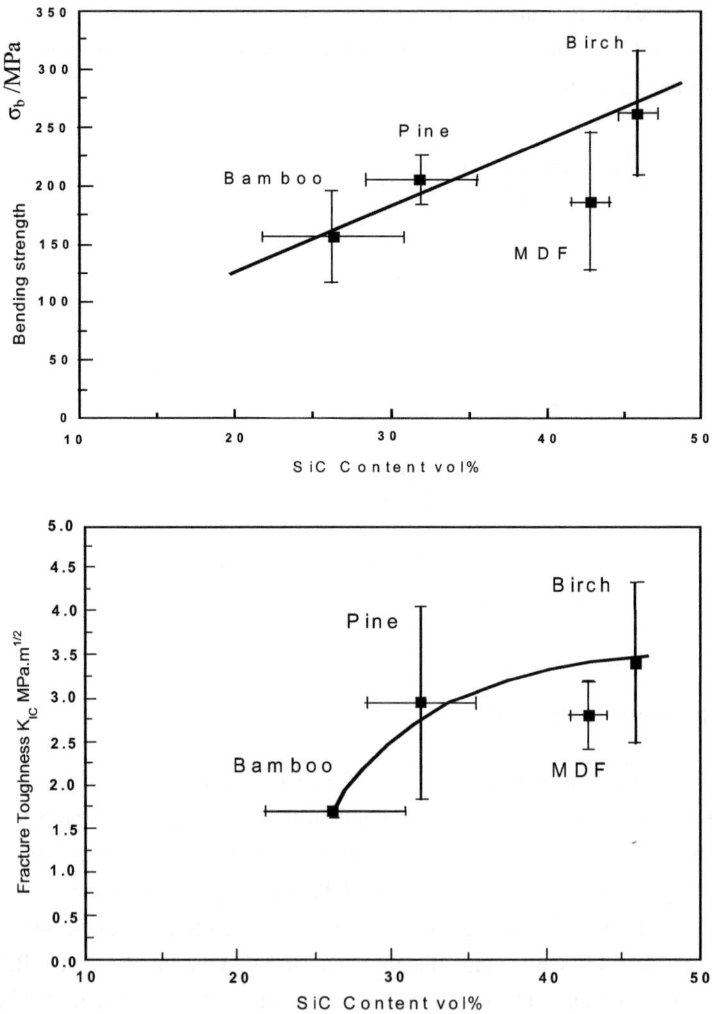

Fig. 4.5. The strength and toughness of Si/SiC ceramics as a function of SiC content

4.3.4 Reaction Mechanism

In order to clarify the reaction mechanism between liquid silicon and porous carbon derived from wood, pine woodceramic and silicon were heated at 1550°C for different holding times, as shown in Figure 4.6. Pine was selected as the model specimen because it has a representative cell structure. It can be seen that .

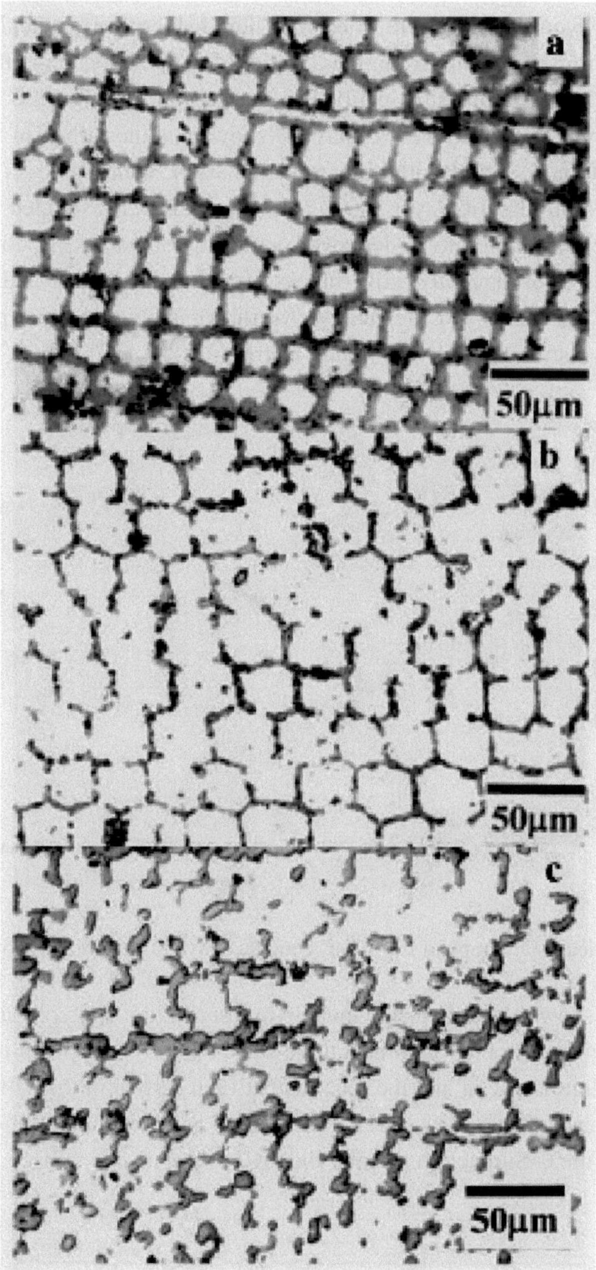

Fig. 4.6. Optical micrograph of Si/SiC ceramics made from pine at 1550°C for (a) 5 min, (b) 15 min and (c) 30 min

the pores of woodceramic were filled with liquid silicon very quickly and densification was finished within 5 minutes. In the initial reaction stage the carbon cell wall reacted with liquid silicon filled in the cell and formed SiC. The reticular framework was preserved well. Figure 4.7 shows that the carbon in the core of the cell-wall has not reacted with silicon at this stage, which gives a dark color in comparison with the SiC layer. As the reaction time was extended up to 15 minutes, the cellular structure is destroyed with the company of C-Si reaction (see Figure 4.6b). Subsequently, SiC blocks were recrystallized and the reticular framework was destroyed drastically. As shown in Figure 4.6c, polygonal SiC grains were distributed in a free silicon matrix.

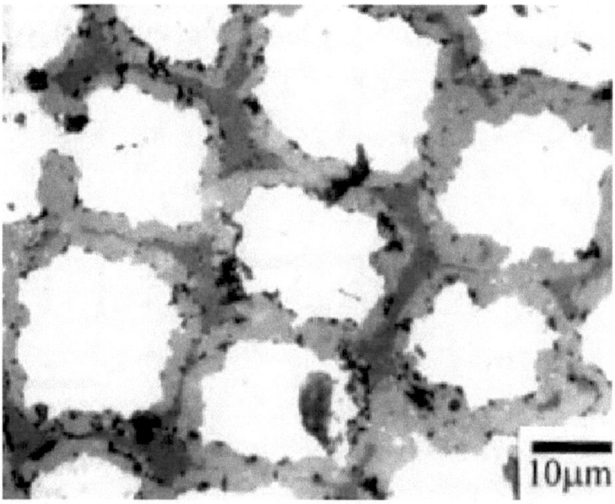

Fig. 4.7. Optical micrograph of the Si/SiC ceramic derived from pine at 1550°C for 5 min

Figure 4.8 shows the schematic model to describe the charcoal-silicon reaction process, which can be divided into three stages. Firstly, the pores in charcoal are filled with liquid silicon and the reaction between carbon in the cell-wall and silicon takes place. Then the cellular structure of SiC begins to fragment with the progress of the C-Si reaction. Finally the SiC blocks are crystallized.

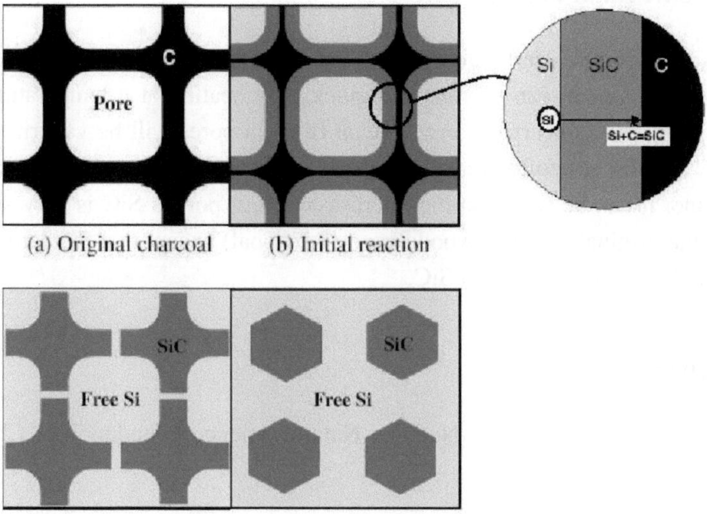

Fig. 4.8. The reaction model for the formation of Si/SiC ceramics

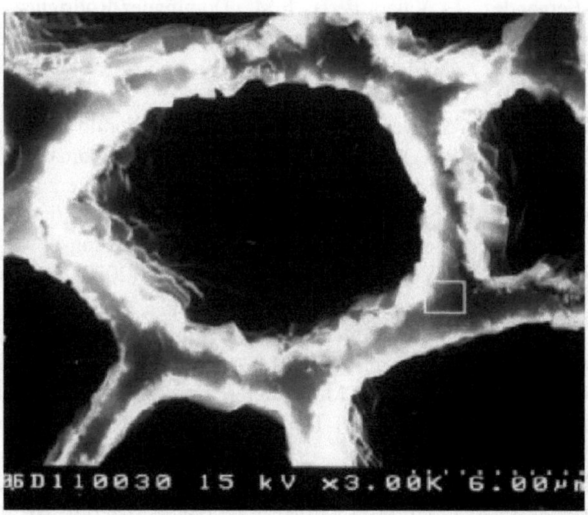

Fig. 4.9. SEM micrograph of porous SiC derived from pine

4.3.5 Removal of Free Silicon

Porous SiC ceramic can be obtained by means of the removal of free silicon. The charcoal and silicon can be removed quickly by heating at a temperature above 1700°C in a vacuum furnace. Free silicon filled in pores will be vaporized before the SiC cellular structure breaks. Figure 4.9 shows the porous SiC ceramic made from pine. It can be seen that the microstructure of porous SiC is very similar to that of the original wood or woodceramic (charcoal) except that the carbon of the cell-wall has been replaced by SiC.

Acknowledgements

This work was supported by National Natural Science Foundation of China (no. 59872025).

References

1. Martinez-Fernandez, J., Valera-Feria, F.M., Dominguez-Rodriguez, A. (2001) Rev Metalurgia, **37**, 295
2. Ashby, M.F. (1989) Acta Metall., **37**, 1273
3. Ashby, M.F. (1991) Acta Metall., **39**, 1025
4. Sieber, H., Kaindl, A., Schwarze, D. (2000) Cfi-Ceram. Forum Int., **77**, 21
5. Okabe, T., Saito, K. (1994) Advanced Materials '93, V/A: Ecomaterials, 681
6. Qiao, G.J., Ma, R., Jin, Z.H. (1997) The 3rd International Conference on Ecomaterials. Japan, 156
7. Singh, M. (2000) Ceramic Engineering and Science Proceedings, 39
8. Martinez-Fernandez, J., Valera-Feria, F.M., Singh, M. (2000) Scripta Materialia, **43**, 813
9. Greil, P., Lifka, T., Kaindl, A. (1998) J. Euro. Ceram. Soc., **18**, 1975
10. Shin, D.W., Park, S.S., Choa, Y.H. (1999) J. Am. Ceram. Soc., **82**, 3251
11. Qiao, G.J., Ma, R., Cai, N. (2002) Mater. Sci. & Eng., **A32**, 301
12. Qiao, G.J., Ma, R., Cai, N. (1999) The 4th International Conference on Ecomaterials. Japan, 301
13. Qiao, G.J., Ma, R., Cai, N. (2002) J. Mater. Process. Tech., **120**, 107
14. Cai, N., Ma, R., Qiao, G.J. (2001) Journal of Inorganic Materials, **16**, 763
15. Cai, N., Qiao, G.J., Jin, Z.H. (2001) Journal of Xi'an Jiaotong University, **35**, 192
16. Greil, P., Lifka, T., Kaindl, A. (1998) J. Euro. Ceram. Soc., **18**, 1961
17. Sieber, H., Hoffmann, C., Kaindl, A. (2000) Adv. Eng. Mater., **2**, 105
18. Sieber, H., Vogli, E., Mullier, F. (2001) Key Eng. Mat., **206—213**, 2013
19. Vogli, E., Mukerji, J., Hoffman, C. (2001) J. Am. Ceram. Soc., **84**, 1236
20. Vogt, U., Herzog, A., Graule, T. (2002) Key Eng. Mar., **206**, 1941
21. Ota, T., Takahashi, M., Hibi, T., (1995) J. Am. Ceram. Soc., **78**, 3409

5 Microstructure through Design in-situ Seeding Technique

J.-L. Huang, H.-H. Lu

5.1 Introduction

Silicon-nitride-based ceramics are attractive for their promising mechanical, thermal and chemical properties. The mechanical behaviors of Si_3N_4 materials are substantially affected by the grain morphology and grain size distribution [1,2]. It was previously discussed that the high fracture toughness observed in silicon nitride ceramics was correlated with their unique microstructure containing an interlocking network of elongated β-Si_3N_4 grains [3,4].

Lange reported that the aspect ratio of β-Si_3N_4 grains in sintered Si_3N_4 increased with the content of α-Si_3N_4 in raw material powders [5,6]. A high α phase content was preferable for the formation of elongated grains in the developed microstructure through a dissolution-reprecipitation process during sintering. Mitomo and Huang showed that a high α content was not necessary for the formation of columnar grains [7]. Their results indicated that β-Si_3N_4 particles in the starting powders could serve as nucleation sites onto which the dissolved α-Si_3N_4 reprecipitated as elongated β-Si_3N_4. Hwang et al [8,9] investigated the nucleation and growth of α' and β'-SiAlON and reported the epitaxial grain growth of α'-SiAlON on α-Si_3N_4. Although the microstructure and defects in seeded Si_3N_4 were previously studied [10,11], there was still not enough microscopy evidence to account for the role of β-Si_3N_4 seed in the nucleation, growth and grain coalescence.

In this study, the microstructure of silicon nitride incorporated with different contents of β-seeds were investigated by image analyzer. The mechanical properties and fracture behavior were also involved and correlated with the grain morphology of sintered specimens.

5.2 Experimental Work

5.2.1 Preparation of β-Si_3N_4 Seeds

Silicon nitride powder was mixed with 6 wt% Y_2O_3 and 2 wt% Al_2O_3, and ball milled in a PE bottle with high purity silicon nitride balls and ethanol for 24 h. The slurry was dried in a rotary evaporator, ground, and followed by screening through a 200-mesh screen. Powders were then heated at 1850°C for 2 h under a nitrogen pressure of 1 MPa to obtain the β-seed. The seed was subsequently pulverized by a mixer mill and screened to pulverize aggregates. Powders were then rinsed in 1 M NaOH solution at 80°C for 1 h to remove the residual glassy phase [7].

5.2.2 Sample Preparation

Si_3N_4 powders were mixed with 6 wt% Y_2O_3, 2 wt% Al_2O_3, and a different amount of seeding in a PE bottle with high purity Si_3N_4 balls and ethanol for 24 h. The slurry was dried in a rotary evaporator, ground, and followed by screening. Powders were then die-pressed at 10 MPa to form a 5 mm × 5 mm × 50 mm green compact and cold isostatically pressed at 100 MPa. Samples were subsequently embedded in a powder bed in a graphite crucible and sintered at different temperatures under 1 MPa N_2 for 1 h.

5.2.3 Microstructural Analysis

Scanning electron microscopy (SEM) was used to examine the microstructure of sintered Si_3N_4. Samples for SEM analysis were cut, ground and polished to 1 μm. Micrographs were revealed by plasma etching with CF_4 and O_2 source at 220 W for 1.5 min. Specimens were chemically etched by molten NaOH at 350°C for 1.5 min followed by Au coating before conducting image analysis.

Foils for transmission electron microscopy (TEM) were ultrasonically machined from sintered bars, dimpled to 25 μm and ion milled by argon ions before scrutiny. A thin layer of carbon was evaporated onto these foils to avoid surface charging under the electron beam.

5.2.4 Characterization

Flexural strength was measured by four-point bend test on a universal testing machine. Fracture toughness was determined by the single-edge precracked beam method [7]. The test procedure and sample preparation were similar to a four-point bend test. The outer and inner spans were 40 and 20 mm, respectively. Samples were tested at a loading rate of 0.5 mm/min. An initial precrack of 1 mm in depth was cut using a diamond blade. Each data point represents an average of six tests.

5.3 Results and Discussion

The TEM image and diffraction pattern of as-heated β-Si_3N_4 seeding are shown in Fig. 5.1. A glassy phase surrounding β-Si_3N_4 particles was detected as indicated by Fig. 5.1b. TEM image and diffraction pattern of the rinsed β-Si_3N_4 seeding are shown in Fig. 5.2. Diffraction pattern shown in Fig. 5.2b clearly indicated that only β-Si_3N_4 and no glassy phase was detected in the rinsed β-Si_3N_4 seeding in comparison with Fig. 5.1. The results of EDS analysis in as-sintered and rinsed β-seeds indicated that most of the aluminum and all of the yttrium in as-heated β-seeds were removed after rinse, suggesting that the glassy phase was effectively removed.

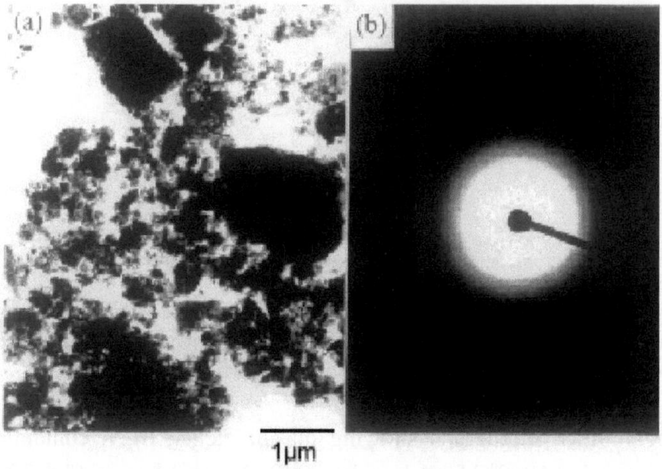

Fig. 5.1. TEM image of as-heated β-Si_3N_4 seeding. Powder was fired at 1850°C for 2 h. (a) The image of as-heated seeding and (b) the diffraction pattern of the glassy region

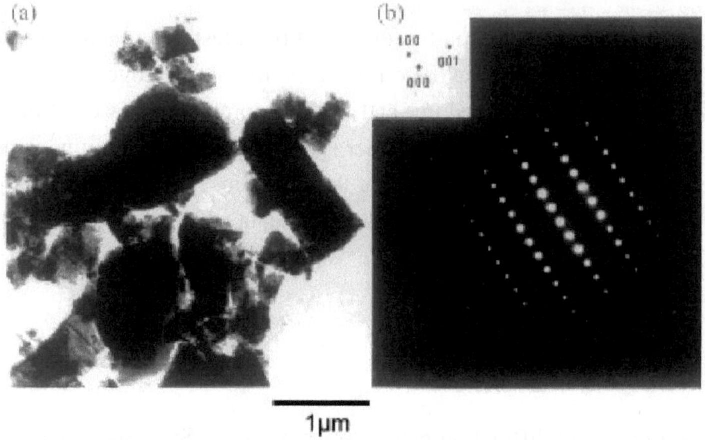

Fig. 5.2. TEM image of rinsed β-Si_3N_4 seeding. Powder was heated at 1850°C for 2 h and washed by 1M NaOH at 80°C for 1 h. (a) The image of rinsed seeding and (b) the diffraction pattern of the region

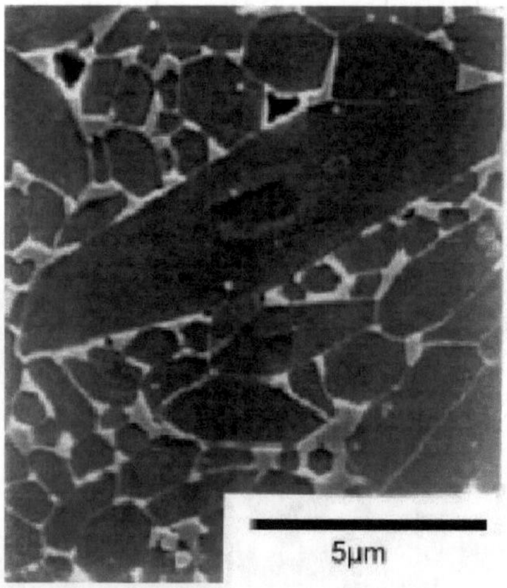

Fig. 5.3. Typical SEM micrograph of silicon nitride with 2 wt% β-seed. Samples were sintered at 1950°C for 1 h under 1 MPa N_2

A typical SEM micrograph showing plasma-etched Si_3N_4 containing 2 wt% β-seeds, sintered at 1950°C for 1 h under 1 MPa N_2, is shown in Fig. 5.3. A typical core/shell structure, both identified as β-Si_3N_4, was observed in large elongated grains but not in small matrix grains. The core was most probably a β-seed which acted as a nucleus during the liquid phase sintering according to the elemental analysis [7,10]. Since it is easier for the new β phase to precipitate on the original β-seed than to nucleate a new β-nucleus [12], more large grains could be developed in samples containing β-seed.

A high-resolution TEM image of the core/shell interface is shown in Fig. 4. The interface indicated by the arrows between the core and shell appeared coherent, as revealed by the continuity of lattice planes across the interface (1010). This suggested an epitaxial growth started from the β-Si_3N_4 seed, and then eventually resulting in the core and shell structure.

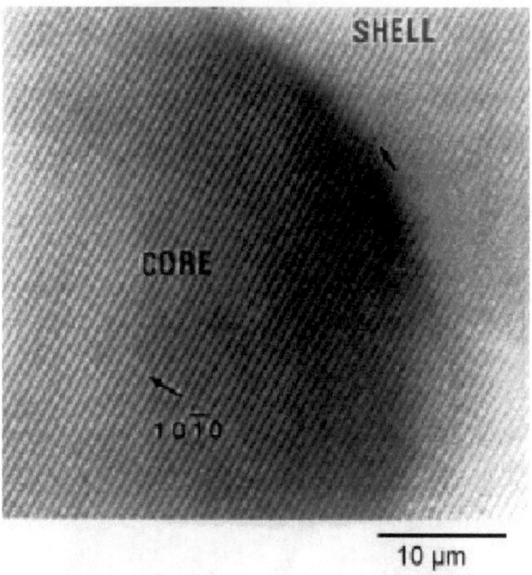

Fig. 5.4. High-resolution lattice image showing the interface between the core and shell

The flexural strength and toughness of the samples containing 0, 2, and 5 wt% β-seeds, sintered at 1950°C for 1 h under 1 MPa N_2, are shown in Fig. 5.5. The samples without β-seeds have a flexural strength slightly higher than the samples with β-seeds. This might be related to the grain size of the samples. The fracture toughness of the samples with β-seeds is always higher than that of the samples without β-seeds. Several mechanisms have been proposed regarding the fracture toughness of self-reinforcement of silicon nitride with large elongated grains or silicon nitride composites [1,13]. The fracture toughness can be correlated to with their microstructure. Crack bridging is proposed to be the mechanism responsible for the toughening effects from the elongated grains of Si_3N_4.

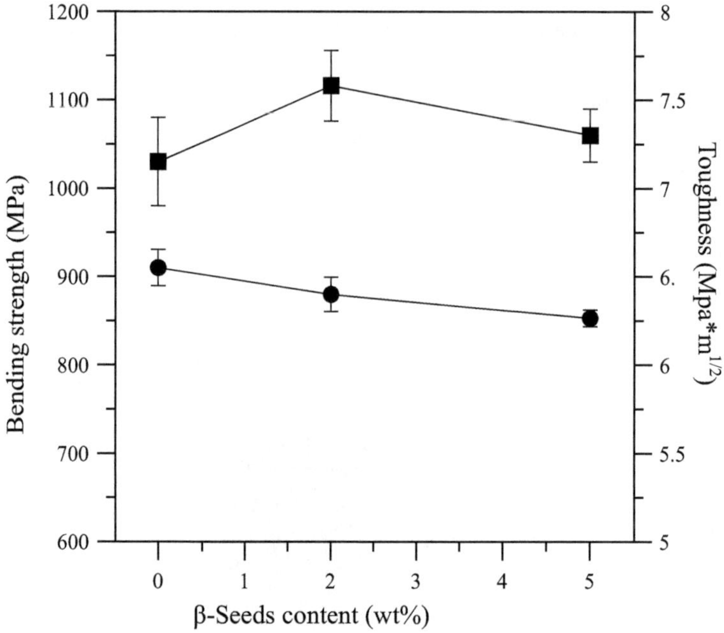

Fig. 5.5. Bending strength and fracture toughness of silicon nitride versus β-seeds content. Samples were sintered at 1950°C for 1 h under 1 MPa N_2

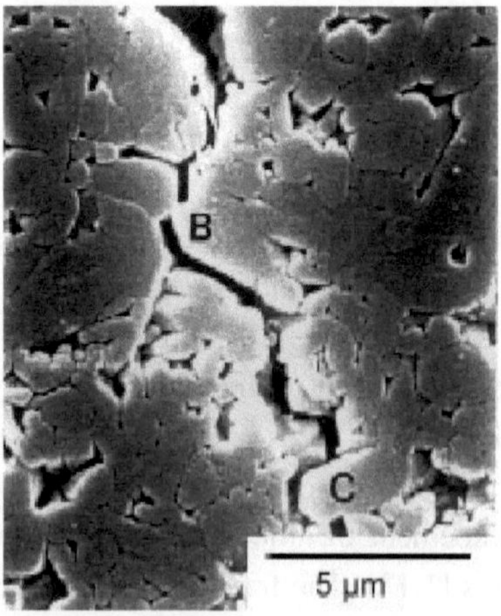

Fig. 5.6. Crack propagating path indicated B: crack deflection, C: grain bridging. Sample containing 2 wt% β-seed was sintered at 1950°C for 1 h under 1 MPa N_2

In order to examine the crack behaviors, a crack propagation path was introduced by a Vickers indentor at 294N on the polished surface of samples containing 2 wt% β-seed (Fig. 5.6). A typical SEM micrograph revealed rodlike morphology of β-Si_3N_4 embedded in fine matrix grains. Both crack deflection and gain bridging were observed along the crack propagation path as indicated in Fig. 5.6.

5.4 Summary and Conclusions

1. A core/shell structure and complicated dislocation image were observed in β-seeded samples owing to the difference in compositions of Si_3N_4 grains.
2. The grain size had a tendency to increase with the incorporation of β-seed. The subsequent precipitation of β-phase Si_3N_4 on the β-seeds is epitaxial growth. The abnormal grain growth occurred in the final stage of sintering.
3. The dominant toughening mechanism with respect to the elongated grains is a crack wake process. The fracture toughness was increased effectively by incorporating β-seeds, without sacrificing the bending strength.

Acknowledgements

The authors would like to thank the National Science Council for providing financial support under the grant NSC 89-2216-E-006-071.

References

1. Becher, P. (1991) J. Am. Ceram. Soc., **74**, 255
2. Li, C-W, Lui, S-C, Goldacker, J. (1995) J. Am. Ceram. Soc., **78**, 449
3. Sajgalik, P., Dusza, J., Hoffmann, M.J. (1995) J. Am. Ceram. Soc., **78**, 2619
4. Hirosaki, N., Akimune, Y., Mitomo, M. (1993) J. Am. Ceram. Soc., 76, 1892
5. Lange, F.F. (1979) J. Am. Ceram. Soc., **62**, 428
6. Lange, F.F. (1973) J. Am. Ceram. Soc., **56**, 518
7. Lu, H-H, Huang, J-L (1999) J. Mater. Res., **14**, 2966
8. Hwang, S-L, Chen, I-W (1994) J. Am. Ceram. Soc., **77**, 1711
9. Hwang, S-L, Chen, I-W (1994) J. Am. Ceram. Soc., **77**, 1719
10. Hirao, J., Nagaoka, T., Brito, M.E., Kanzaki, S. (1994) J. Am. Ceram. Soc., **77**, 1857
11. Ming, X.G., Wilkinson, D.S., Weatherly, G.C., Ye, H.Q. (1997) J. Mater. Sci., **32**, 1431
12. Lee, R. R-R, Chen, C-J, Lin, L-T. (1994) Silicon-Based Structural Ceramics, Ceram. Trans. **42,** ed. by Sheldon, B.W., Danforth, S.C., The American Ceramic Society, Westerville, Ohio, 221
13. Faber, K.T., Evans, A.G. (1983) Acta Metall., **31**, 565

6 Directional Solidification : ZrB_2-LaB_6 Eutectic Composites

L. Zhang, Y. Xu, L. Cheng, C. Chen

6.1 Introduction

The composites prepared by in-situ directional solidification have received considerable attention because of their unique microstructure. In recent decades, a great effort has been directed towards the development of oxide-oxide [1] and metal-metal [2] eutectic composites, and a few carbide-diboride eutectic systems [3]. However, little work has been done on boride-boride eutectics, which usually exhibit high melting points. Recently, the d- and f-transition metal-boride eutectics have been studied by Paderno [4], and unusual thermal and mechanical properties were reported. The LaB_6-ZrB_2 system was reported to have superb fracture toughness and strength. The electrical conductivity of the boride composites is also high. It suggests that the boride composites are potential materials for structural and functional applications.

The ZrB_2-LaB_6 system has an eutectic temperature of about 2470°C. Compared with the sintered borides, the toughness and density of eutectic ZrB_2-LaB_6 are better. The aim of this study was to investigate the microstructure and its stability at elevated temperatures. Mechanical properties such as the hardness and fracture toughness were also determined.

6.2 Experimental Work

The LaB_6 and ZrB_2 powders used in the experiment were produced by the Institute Problem of Materials Science in Ukraine. The purity of LaB_6 was >99.4%, with major impurities of 0.3% Fe and 0.1% O, and with minor impurities of Si and Mg. The purity of ZrB_2 was >98.2%, with major impurities of 0.3% Fe, 0.3% Mg and 0.15% O, and minor impurities of 0.03% Ca and 0.04% Al. The impurity concentration was reported by the supplier. The borides with the eutectic composition (20.5 wt% ZrB_2-79.5 wt%LaB_6) were thoroughly mixed, dry-pressed, and melted with an electric arc to form rods of 6 mm in diameter and 80 mm in length. The rods were then directionally solidified to form fiber/matrix composites through zone-melting in an electric arc furnace.

The final specimens were cut by electric discharging along and across the axes of the rods to obtain the longitudinal and transverse sections. After polishing with diamond paste and etching with nitric acid, the specimens were examined by an ARMAY Model 1000B scanning electron microscope (SEM). The line-intercept method was used to measure the inter-fiber spacing. Thin foils were prepared by ion beam thinning and examined by JEM-200CX transmission electron microscope (TEM).

The oxidation properties of the composites were examined by measuring the weight change using a TGA. The samples for the oxidation test were cut from the directionally solidified rod. By changing the ratio of oxygen over argon, the oxygen partial pressure during the oxidation test environment was carefully controlled.

Microstructure stability was evaluated by comparing the microstructure of the composite before and after heat-treatment at 1650°C for 90 hours.

6.3 Results and Discussion

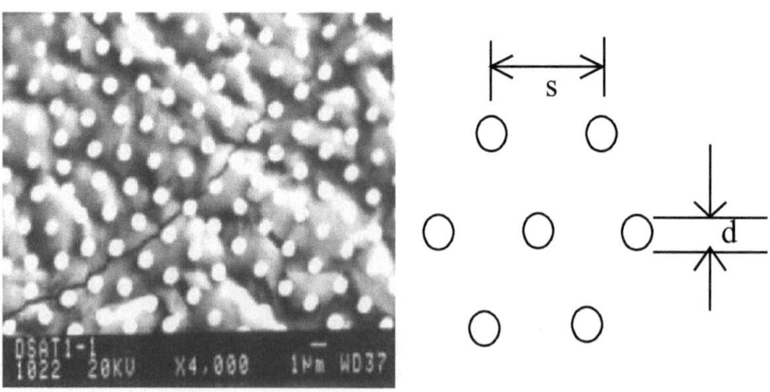

(a) (b)

Fig. 6.1. Microstructure (a) and idealized fiber array (b)

The directionally solidified LaB_6-ZrB_2 eutectic composite has a unique microstructure, with the ZrB_2 fiber distributed uniformly within the LaB_6 matrix, as shown in Fig. 6.1a. Table 1 gives the solidification rate, fiber diameter and fiber spacing. The fiber spacing was derived from the volume fraction of fiber by assuming a fiber array as shown in Fig. 6.1b. The volume fraction of fiber was

Table 6.1. Fiber diameter and distance in the LaB_6-ZrB_2 composite at different solidification rates

Solidification rate [mm/min]	Fiber diameter, d [μm]	Fiber distance, S [μm]
0.5	1.2	2.8
1.0	1.0	2.3
1.5	0.9	2.1
5.0	0.3	0.7
8.0	0.2	0.5

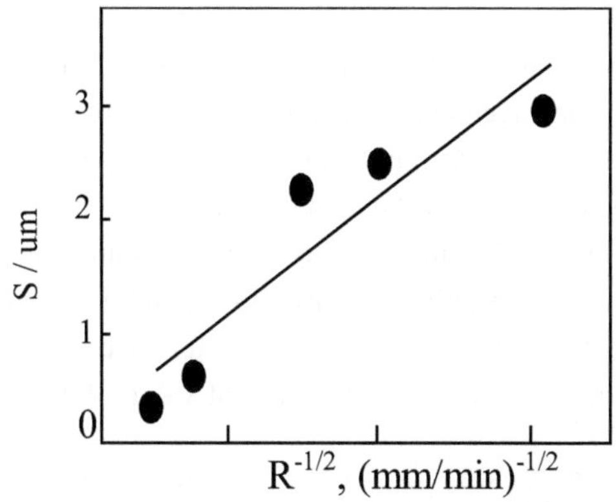

Fig. 6.2. Relationship between fiber spacing, S, and solidification rate, R

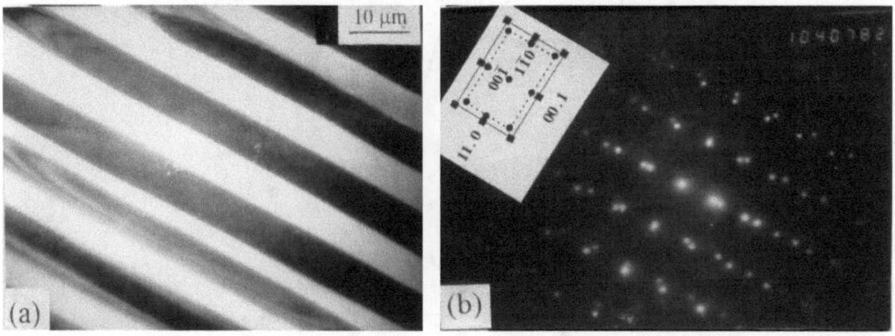

Fig. 6.3. (a) Transmission electron micrograph and (b) selected electron diffraction for the longitudinal section of LaB_6-ZrB_2 eutectic

calculated as 17% from the weight percentages and the densities of the two components. Fig. 6.2 shows that the fiber spacing is proportional to the square root of the solidification rate, a relationship suggested by Jackson and Hunt [5].

Figure 6.3 shows a transmission electron micrograph and an electron diffraction pattern taken from a longitudinal section of the eutectic ingot at a growth rate of 2.5 mm/min. The fibers were well orientated along the growth direction and the interfaces are smooth. This suggests that low energy interface is formed in this system. The growth direction of the LaB_6 matrix and the ZrB_2 fibers were $[001]_{LaB6}$ and $[001]_{ZrB2}$ directions respectively, which were perpendicular to $(001)_{LaB6}$ plane. The crystallographic relation between the matrix and the fiber deduced from Fig. 6.3b was:

growth direction: $[001]_{LaB6}//[001]_{ZrB2}$

interfacial plane: $(110)_{LaB6}//(010)_{ZrB2}$

Figure 6.4 shows the interfacial structural relationship between LaB_6 and ZrB_2 phases. It is known that lanthanum hexaboride has a cubic CaB_6-type structure with a single cubic arrangement of metal atoms centered by boron atom octahedron. The zirconium diboride has hexagonal AlB_2-type structure with alternate crystallographic plane along the hexagonal c-axis, which has flat hexagonal nets of metal or boron atoms. The intraoctahedral B-B distance in lanthanum hexaboride and B-B distance in zirconium diboride in the flat hexagonal net is 0.1765 nm and 0.1829 nm respectively. The lattice mismatch between the matrix and the fiber is only 3.16%, which is well below the critical value of 16% for a semicoherent interface [1].

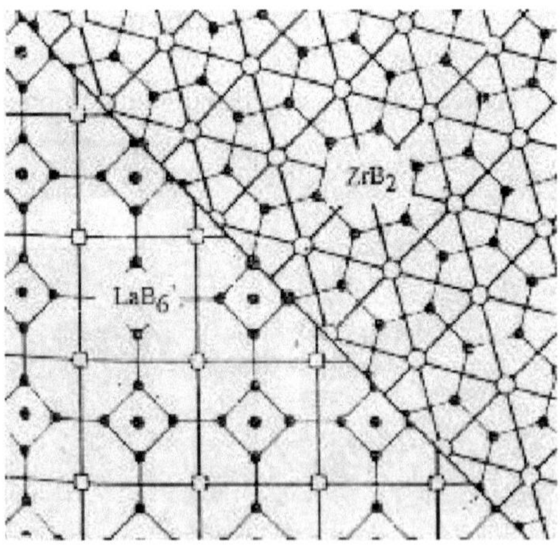

Fig. 6.4. Structural relationship between LaB_6 and ZrB_2 phases during co-crystallization

Figure 6.5 and Table 6.2 give the hardness and fracture toughness of the composite. The results indicate that the fracture toughness is highest if the flaw propagates in direction A. This is mainly induced by the fiber reinforcement. In this case, it will be easier for the flaw to change its direction and propagate along

Table 6.2. Hardness and fracture toughness of LaB_6-ZrB_2 composites for different flaw directions

Direction of flaw propagation	A	B	C
Hv [GPa]	31.4	31.4	28.8
K_{IC} [$MPa \cdot m^{1/2}$]	17.8	8.2	8.7

the fiber/matrix interface. Accordingly, the fracture toughness was increased because of the fiber pullout. Flaw propagation in direction B is the easiest because the direction of the flaw propagation is the same as the direction of the fiber/matrix interface. If a flaw propagates in direction C, it will be deflected around the fiber as indicated with white arrows in Fig. 6.5c. This will slightly increase the fracture toughness compared with the case of direction B.

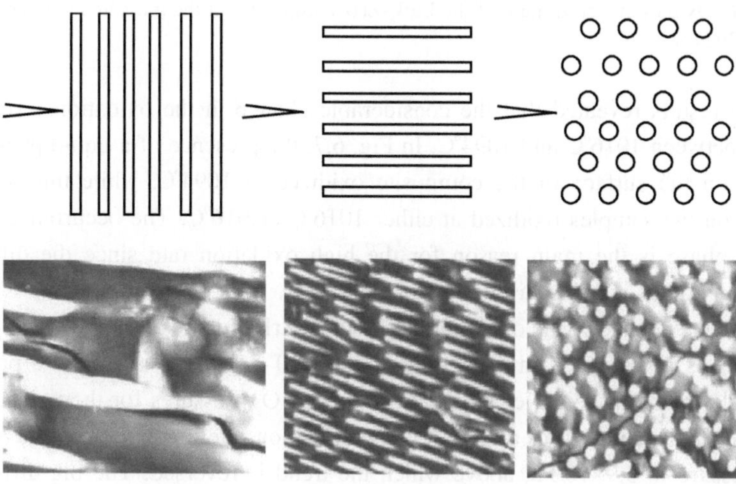

A B C

Fig. 6.5. Propagation directions of flaws in the composites

Figure 6.6 shows the oxidation behavior of the composite at different temperatures. It can be seen that the oxidation weight gain is very large at a temperature of 1094°C. However, the oxidation weight gain of the composite does not change significantly as the temperature increases from 916°C to 1016°C.

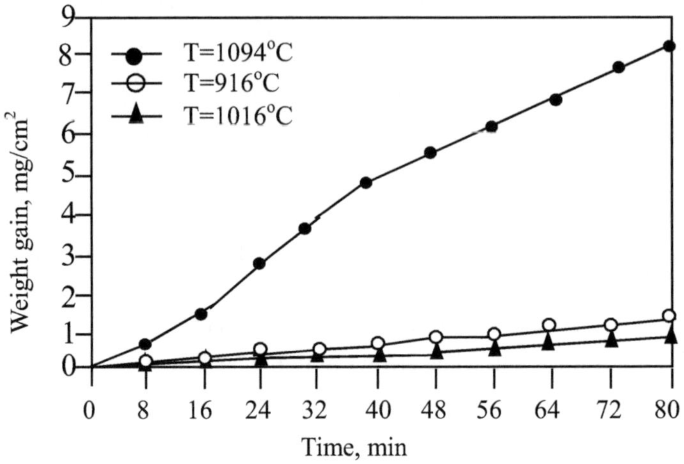

Fig. 6.6. Oxidation weight gain of the LaB_6-ZrB_2 composites at the indicated temperatures (P_{O2} = 292 torr)

The results revealed that the considerable change in the oxidation layer took place between 1016°C and 1094°C. In Fig. 6.7, the presence of a liquid phase was found on the surface of the composite oxidized at 1094°C, while this was not found on the samples oxidized at either 1016°C or 916°C. The occurrence of the liquid phase is the main reason for the high oxidation rate since the diffusion speed in liquids is much faster than that in solids.

The dependence of the oxidation of the $LaB_6{\cdot}ZrB_2$ composites on oxygen partial pressure is investigated in a P_{O2} range of 1.5×10^4—7.2×10^4Pa. Figure 8 presents parabolic plots of weight-change data at various O_2 pressures for the oxidation of the $LaB_6{\cdot}ZrB_2$ composites at 1094°C. The oxidation rate increases with increasing O_2 pressure to 3.9×10^4Pa, above which the trend is reversed. The big difference between the weight changes when the oxygen pressure is raised from 1.5×10^4 to 3.9×10^4Pa is due to the formation of large amounts of oxides and the concurrent evaporation of B_2O_3. The oxidation rate decreases with the increase of P_{O2} in a range of 3.9×10^4—7.2×10^4Pa, which may be due to rapid active oxidation of ZrB_2 in this condition. The resulting ZrO_2 provides more efficient oxidation resistance than that provided by the lanthanum oxide or its compounds.

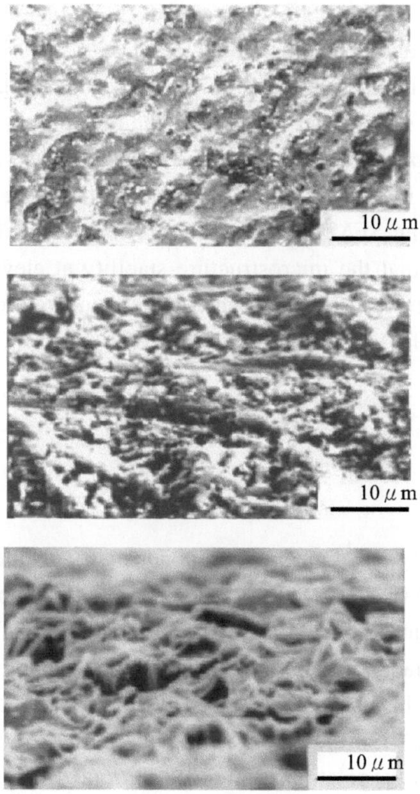

(a) at 916°C

(b) at 1016°C

(c) at 1094°C

Fig. 6.7. Surfaces of the composites oxidized at the indicated temperatures

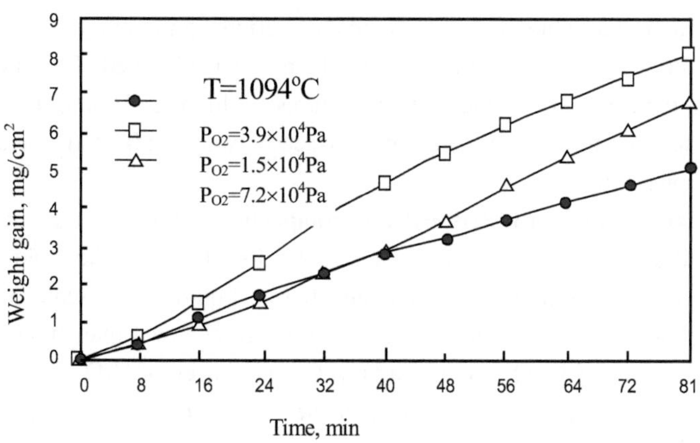

Fig. 6.8. Oxidation weight gain of the LaB_6-ZrB_2 composite at different oxygen pressures (T = 1094°C)

The products in the oxidation layer are determined by X-ray analysis as $La_2O_3{\cdot}B_2O_3$, $La_2O_3{\cdot}3B_2O_3$, ZrO_2, and a small amount of glass phase. The boron much less in the oxidation layer is much less than in the bulk, indicating a loss of boron during the oxidation process. In order to evaluate the microstructure stability of the composite at high temperatures, a composite was heat-treated at 1650°C for 10, 50 and 90 hours and the microstructure of the composite before and after heat-treatment was observed using SEM. No detectable difference of the microstructure was found, indicating that the microstructure stability at elevated temperature is excellent. This is mainly due to the high eutectic temperature between LaB_6 matrix and ZrB_2 fiber (2470°C). It may also be affected by the semi-coherent interface.

6.4 Conclusions

Directional solidification of LaB_6-ZrB_2 eutectic was conducted with the electron arc floating zone melting method (EAFZM). The LaB_6-ZrB_2 in-situ composite was obtained with this method, and an orientated microstructure with ZrB_2 fibers uniformly distributed in the LaB_6 matrix was obtained. The range of diameter of the ZrB_2 fibers was ~0.2—1.2 μm, with fiber lengths up to 100 μm. The fiber number per unit area can be up to 108/cm^2. For the ingots with planar growth, both phases grow in the [001] directions with the relationship (110)LaB_6//(110)ZrB_2 observed. The misfit between the two phases was calculated to be below the critical value of 16% for a semi-coherent interface.

The fracture toughness of the composites varied significantly in different directions. A high fracture toughness of 17 MPa·m$^{1/2}$ was obtained. The microstructure the composite is very stable at temperatures as high as 1650°C because of the eutectic temperature is high and the interface between LaB_6 matrix and ZrB_2 fiber is semi-coherent.

The oxidation of LaB_6-ZrB_2 eutectic composites follows a parabolic law below 1094°C. It exhibits paralinear behavior above 1094°C due to the rapid evaporation of B_2O_3 and the formation of lanthanum borate liquid in the scale, which offers less of a barrier to oxygen diffusion. At temperatures lower than 1094°C, the scale consists mainly of $La_2O_3{\cdot}3B_2O_3$, $La_2O_3{\cdot}B_2O_3$ and traces of ZrO_2. The oxidation is controlled by oxygen diffusion through the scale. The oxidation of LaB_6 seems more severe at lower oxygen pressure. This can be attributed to the formation of two layers with ZrO_2 deficiency in the outer oxidized layer.

Acknowledgements

The authors thank the support of the Chinese National Foundation for Natural Sciences under Contract No. 59582008 and the Chinese Defense Foundation for Sciences under Contract No. 96J12.1.6.

References

1. Minford, W.J., Bradt, R.C., Stubican, V.S. (1979) J. Am. Ceram. Soc., **62**, 154
2. Hertzberg, W., Kraft, K.W. (1963) Trans. Met. Soc. AIME, **227**, 580
3. Sorrell, C., Beratan, H.R., Bradt, R.C., Stubican, V.S. (1984) J. Am. Ceram. Soc., **67**, 190
4. Paderno, Yu.B., Paderno, V.N., Filippov, V.B. (1991) AIP Conf. Proc., **231**, 561, AIP, New York
5. Jackson, K.A., Hunt, J.D. (1966) Trans. Metall. Soc. AIME, **236**, 1129

Acknowledgements

The author[illegible] thank the support of the Chinese Na[illegible] Science under [illegible] and [illegible] [illegible].

References

[illegible]

7 A Co-Sedimentation Technique to Fabricate Continuous Gradient Composites

L. Zhang, Z. Yang, Q. Shen

7.1 Introduction

Functionally graded materials (FGMs) are a class of composite materials that exhibit inhomogeneous distributions of different components on the scale of the entire material. Because of its properties, FGM can be used for engineering applications where mono-component materials or homogeneous composites fail. Processing techniques used to produce FGM include solidification processing, chemical vapor deposition, powder metallurgy, and co-sedimentation. Among these techniques, the most promising one for the fabrication of large products with continuous and smooth variations in composition and in microstructure is co-sedimentation. The process involves the selection of raw material powders with appropriate sizes and ratios, followed by the segregation of particles and liquid in a gravitational field to create a desired gradient in suspension, and finally the solidification of the spatially graded structure. Various FGMs, such as metal-ceramics and metal-alloy-ceramics, have been successfully synthesized by this method [1—13]. However, the technology remains in its experimental stage, mainly due to a lack of theoretical study on this method.

The objective of the present work is to set up a model that reveals the relations between the compositional distribution of deposited body and the powder characteristics of raw materials as well as the settling parameters.

In the present study, the "powder characteristics" refers to the density and the particle size distribution of raw material powder. Theoretical study on co-sedimentation is separated into two parts, prediction and design. Prediction refers to the calculation of the compositional distribution in FGM based on the powder characteristics and settling parameters, whereas design refers to the computation of the powder characteristics and settling parameters based on the compositional distribution in FGM. Here we confine our discussions to the prediction element.

7.2 Physical Modeling

As seen in our daily lives, in a stationary container filled with a suspension consisting of a liquid and some suspended particles, the particles may settle through the liquid and pack on the bottom of the container under the influence of gravity. When carefully observing, one can find that the large particles settle faster than the small particles. As a result, there is a separation between particles due to different settling velocities. However, the case is not the same within the whole container. There is a small region, near the bottom of the container, in which the particles are scarcely separated due to a far shorter settling distance. Only when a separation between particles is achieved during particulate sedimentation can a gradient of particle size or component be formed throughout the thickness of sediment body. If an additional settling distance is added under the suspension, that is, particles begin to settle at a height of liquid, the problem existing in the conventional settling process mentioned above can be solved. A simplified physical model is shown in Fig. 7.1.

7.3 Numerical Modeling

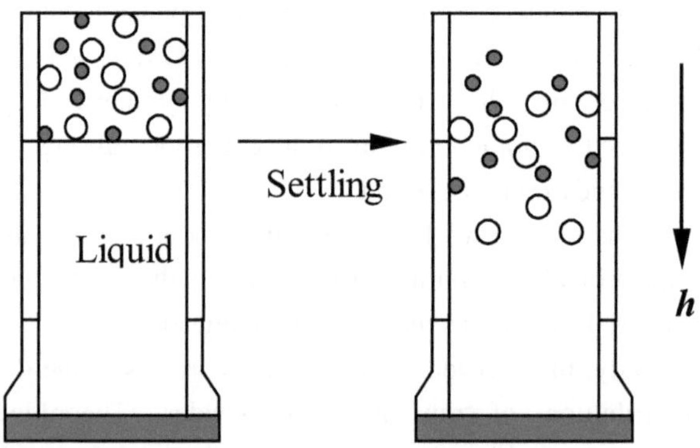

Fig. 7.1. Schematic illustration of particulate sedimentation

The cylindrical vessel filled with a suspension has a height of h and a cross-section area of S. The following assumptions are made in the present model:

1. Particles of each component and liquid are treated as separate phases.
2. Particles are rigid and spherical and perfectly dispersed (no coagulation in suspension).

3. The motion of particles is slow and reaches the terminal velocity instantaneously, i.e. in the laminar flow region, the vertical motion of particles under gravity is predominant and stable.
4. No interactions occur between particles.
5. Volume change during settling is negligible.
6. The sediment height is far less than the suspension height.

With these assumptions, a working model can be formulated as follows.

7.3.1 The Continuum Equation

The continuum equation for particles is:

$$\frac{\partial C(h,t)}{\partial t} + \frac{\partial (UC(h,t))}{\partial h} = 0, \tag{7.1}$$

Where C is the local concentration of particles, U is the settling velocity of particles, which is unrelated to the settling height h. So (7.1) is converted to the form

$$\frac{\partial C(h,t)}{\partial t} + U\frac{\partial C(h,t)}{\partial h} = 0 \tag{7.2}$$

From Stokes law, the settling velocity of particles under gravity in the laminar flow region is

$$U = \frac{D^2 g\left(\rho_{\text{Particle}} - \rho_{\text{Liquid}}\right)}{18\eta}, \tag{7.3}$$

Where D is the equivalent Stokes diameter of particle, g is the standard gravity acceleration; η, is the viscosity of suspension, ρ_{Particle} and ρ_{Liquid} are the densities of particles and liquid respectively.

7.3.2 Initial Conditions

At time t=0, the particles' concentration in suspension is uniform.

$$C(h,0) = \begin{cases} C_0 = \dfrac{m_p}{V_p + V_l} & 0 \le h \le R \\ 0 & R < h \le H \end{cases} \tag{7.4}$$

where V_p and V_l are the volumes of particles and liquid in initial suspension separately, m_p is the mass of particles, R is the height of initial suspension.

7.3.3 Mathematical Model Based on Co-sedimentation

After a period of time t, the particles' concentration at a height h in suspension is

$$C(h,t)= \frac{m'_p}{V'_p + V_l} \tag{7.5}$$

where m'_p is the total mass of raw particles, which have the same sizes as the ones suspended at a height h, and V'_p is the corresponding total volume. Because the particles' concentration is very low and the volume of particles is much smaller than that of liquid, combining (7.4) and (7.5) yields

$$\frac{C(h,t)}{C_0} = \frac{m'_p}{m_p} = \begin{cases} \dfrac{\int_{D_{\min}}^{D_{t1}} f(D)dD}{\int_{D_{\min}}^{D_{\max}} f(D)dD} & 0 \le h \le R \\ \\ \dfrac{\int_{D_{t2}}^{D_{t1}} f(D)dD}{\int_{D_{\min}}^{D_{\max}} f(D)dD} & R < h \le H \end{cases} \tag{7.6}$$

$$\mathtt{1}_{t1}\mathtt{1}\frac{h}{t} \tag{7.7a}$$

$$\mathtt{1}_{t2}\mathtt{1}\frac{h-R}{t} \tag{7.7b}$$

where $D_{\max}$ and $D_{\min}$ are the maximum and minimum diameters of a raw powder respectively, and $f(D)$ is the frequency mass distribution. Because $\int_{D_{\min}}^{D_{\max}} f(D)\mathrm{d}D = 1$, (7.6) can be written as follows

$$C(h,t)= \begin{cases} C_0 \int_{D_{\min}}^{D_{t1}} f(D)\mathrm{d}D & 0 \le h \le R \\ \\ C_0 \int_{D_{t2}}^{D_{t1}} f(D)\mathrm{d}D & R < h \le H \end{cases} \tag{7.8}$$

At a time of t the mass of particles suspended in liquid is

$$M(t)=S\int_0^H C(h,t)\mathrm{d}h \tag{7.9}$$

while the deposited mass is

$$W(t)=m_p - S\int_0^H C(h,t)\mathrm{d}h \tag{7.10}$$

If a suspension includes two types of components A and B, the deposited mass of each component is complant with (7.10). The content of component A in deposited body is

$$Z(t)=\frac{\dfrac{\mathrm{d}W_A(t)}{\mathrm{d}t}}{\left(\dfrac{\mathrm{d}W_A(t)}{\mathrm{d}t}+\dfrac{\mathrm{d}W_B(t)}{\mathrm{d}t}\right)} \tag{7.11}$$

The sediment height at an arbitrary time t is given by

$$L(t)=\frac{\left(\dfrac{W_A(t)}{\rho_A}+\dfrac{W_B(t)}{\rho_B}\right)}{S} \tag{7.12}$$

Equations (7.11) and (7.12) connect the compositional distribution of final product ($Z(t)$, $L(t)$) with the powder characteristics and settling parameters. These equations are solved using one-dimensional, fixed-grid, single-domain numerical solution method through a computer.

7.4 Results and Discussions

7.4.1 The Effect of Particle Sizes on the Deposited Mass per Unit Time

Figure 7.2 shows the deposited mass fraction of Mo particles with time. A pure ethanol with the viscosity of 1.199 cP and the density of 0.789 g/cm^3 is employed as dispersing liquid due to its good wetting for both Mo and Ti powders. Both suspension height and pure liquid height are initially set at 50 cm, and the initial average concentration of suspension is 0.6% [4]. In the early stage of sedimentation, the deposited mass of Mo particles remains zero. After a period of time Mo particles begin to pack on the bottom of a sedimentation tube and the deposited mass fraction continuously varies with time. When a suspension only contains mono-sized particles, the deposited mass fraction linearly changes with time, that is, the deposited mass fraction per unit time (the slope of the deposited mass–time curve) remains constant, as in Fig. 7.2A. Now other particles with a different size are added into the suspension, so that there are two sizes of particles in the suspension. Several turning points then appear in the deposition mass curve, as shown in Fig. 7.2B. Such changes are caused by the change in particle size. During sedimentation, large particles (5μm Mo particles) first pack on the bottom of the sedimentation tube, and subsequently the small particles (4 μm Mo particles) begin to stack on the bottom. After a period of time, large particles completely settle out of the suspension, but the small particles continue settling. So a series of changes of deposited mass appears in the whole sedimentation process. Based on this case, we can derive the fact that when particle sizes continuously change, the deposited mass of particles per unit time also varies continuously. Moreover, from (7.11), the component content in sediment is determined by the deposited mass per unit time. Consequently, the continuous change in particle size results in the continuous change of composition in a deposited body, which is the basis of forming FGM with continuous change of composition by co-sedimentation.

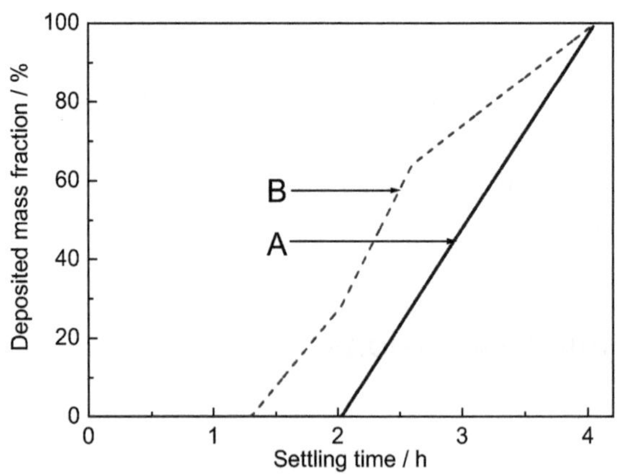

Fig. 7.2. The deposited mass fraction of Mo particles with (A) the diameter of 4 μm and (B) the diameter of 4 and 5 μm

7.4.2 The Influence of Suspension Height

The particle size distributions of Ti and Mo powders are shown in Figs. 7.3. and 7.4 shows the calculated content of Mo along the thickness direction of deposited body. The initial liquid height is fixed at 50 cm. The suspension heights are set to 20 cm and 100 cm, respectively. The calculated the results are shown in Fig. 7.4. We can see that the suspension heights have a strong influence on the component distribution. Such a case is different from that in the conventional settling process [4].

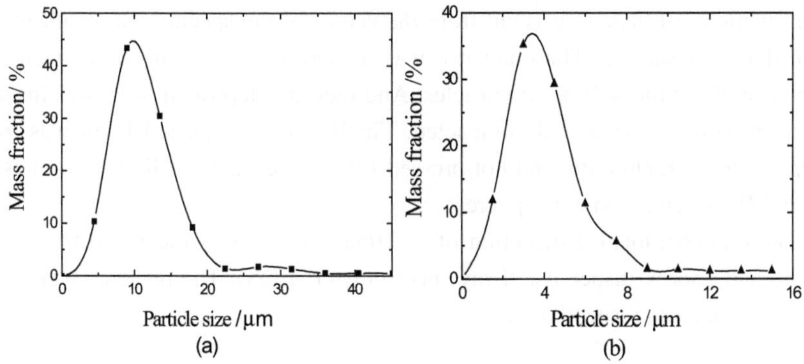

Fig. 7.3. Particle size distribution of (a) Ti and (b) Mo powders

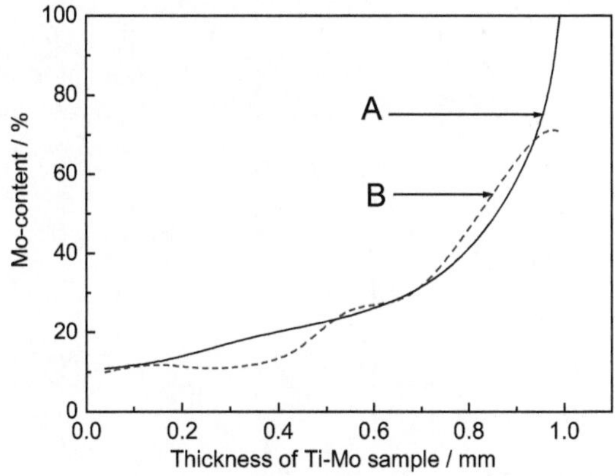

Fig. 7.4. The Mo content calculated along the thickness direction

7.5. Validation of the Model

The Ti and Mo powders (particle size distributions shown in Fig. 7.3) were carefully weighed and subsequently placed in a large beaker containing pure ethanol to make a suspension with a volume concentration of 0.6%. To break up powders into individual particles, the suspension was ultrasonically treated. Then it was poured into a sedimentation tube with a diameter of 45 mm and a height of 1000 mm. Due to the gravity Ti and Mo particles began to settle.

After particles completely settled out of the suspension, the liquid above the deposit body was removed and the deposit was thoroughly dried. The central 45 mm diameter of deposit was cut from the center of the specimen to result in a 32 mm diameter sample. The outer rim was removed, as the wall of sedimentation tube can affect the settling of particles. And then the deposited body was installed in a pressure chamber and compacted. Finally, the compacted body was transferred into a graphite die and hot-pressed for an hour at 1673 K under a pressure of 20 MPa in an argon atmosphere.

The compositional distribution of the final sample was measured by electron probe and energy dispersive X-ray spectrometer (EDS). The results are shown in Figs. 7.5 and 7.6, respectively.

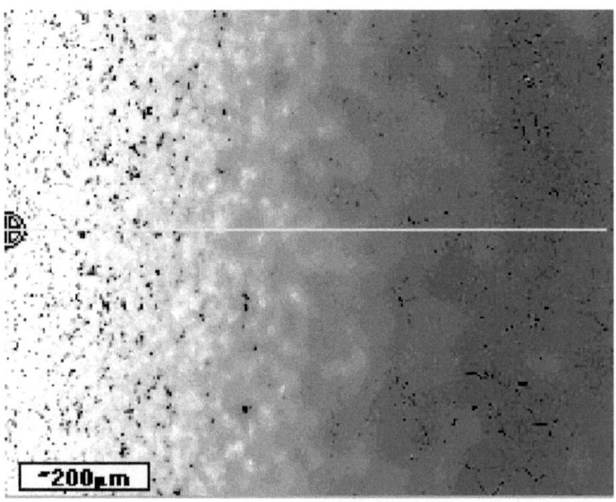

Fig. 7.5. Back-scattered electron image of Ti-Mo gradient materials

Figure 7.5 shows the microstructure of Ti-Mo profiles along its thickness direction. As expected, the continuously graded structures with elements Ti and Mo are formed. Figure 7.6 shows the calculation results and the experimentally measured results. The predicted values are in close agreement with experimental results. It is also noted that the calculation values are slightly lower than the experimental values on the Ti-rich side, in contrast to the Mo-rich side. This phenomenon can be explained as follows. During the settling process, the large particles of Ti first deposited on the bottom of sedimentation tube and formed a loose structure. Some of the following Mo particles filled in the voids of the loose structure, which resulted in a decrease of Mo content in the upper part of the deposit and an increase in the lower part.

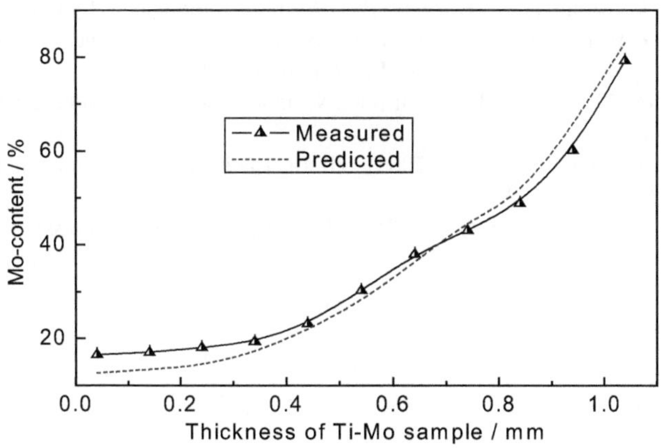

Fig. 7.6. Comparison of the predicted and measured values

7.6 Conclusions

1. A model for forming FGM using co-sedimentation is established. The model can predict the compositional distribution in the final product by inputing the powder characteristics and settling parameters.
2. The size variation of the raw material leads to continuous variations of the composition in the deposited body.
3. Continuous gradient composites can be successfully prepared by using the co-sedimentation technique. The composition profile within the composite can be predicted with the model proposed in the present study.

Acknowledgements

The present study is sponsored by the National Natural Foundation of China under grant no. 50171049.

References

1. Jüngling, T., Bronback, J., Kieback, B. (1994) 3rd International Symposium on Structural and Functional Gradient Materials,. ed. by Ilschner, B., Cherradi, N., Polytechniques et Universitarires Romandes Press, Lausanne, 161—166
2. Huang, J.H., Zhao, J., Li, J.F., Akira, K., Watanabe, R. (1998) Chinese Science Bulletin, **43**, 550
3. Kapelski, G., Varloteaux, A. (2000) 6th International Symposium on Functionally Gradient Materials, USA
4. Yang, Z., Zhang, L., Shen, Q. (2002) J. Mater. Sci. Technol., (in press)

8 Spark Plasma Sintering : A Promising New Technique and its Mechanism

D. Zhang, Z. Fu, R. Yuan, J. Guo

8.1 Introduction

The sintering of ceramic powder compacts is usually carried out by raising the temperature to a point below the melting point of the main component in the compact. Mass transportation, such as diffusion and/or vapor transportation, can take place during sintering. Though the time at the sintering temperature is usually around 1 h, the time needed to heat up and to cool down the sintering furnace is nevertheless long. Sintering is thus a time-consuming process. Furthermore, the sintering profile has to be controlled carefully in order to achieve high density.
An external pressure can enhance the densification rate of a ceramic powder compact. With the help of the external pressure, the maximum temperature needed to achieve full density can be slightly lower than that needed for pressureless sintering. However, the time to heat up and cool down the hot-press furnace is still considerable.

The use of an electric field or current to influence material processing has been attempted for many years. The presence of an electric field can affect the defect migration, crystal growth, evaporation, phase transformation and diffusion, etc. Recent studies demonstrated that applying both external pressure and an electric field simultaneously could significantly enhance the densification of metallic or ceramic powder compacts. Coarsening is not affected by the pressure. The previous studies also showed that a pulsed electric current is more beneficial than a steady one in terms of densification enhancement. Spark plasma sintering (SPS), also known as pulse electric current sintering, is a novel processing technology for materials synthesis and processing.

Figure 8.1 is a schematic sketch of an SPS facility. The SPS has the capability to densify ceramics, metals and composites at a relatively low temperature in a very short time. However, the sintering mechanism for SPS is not well established yet.

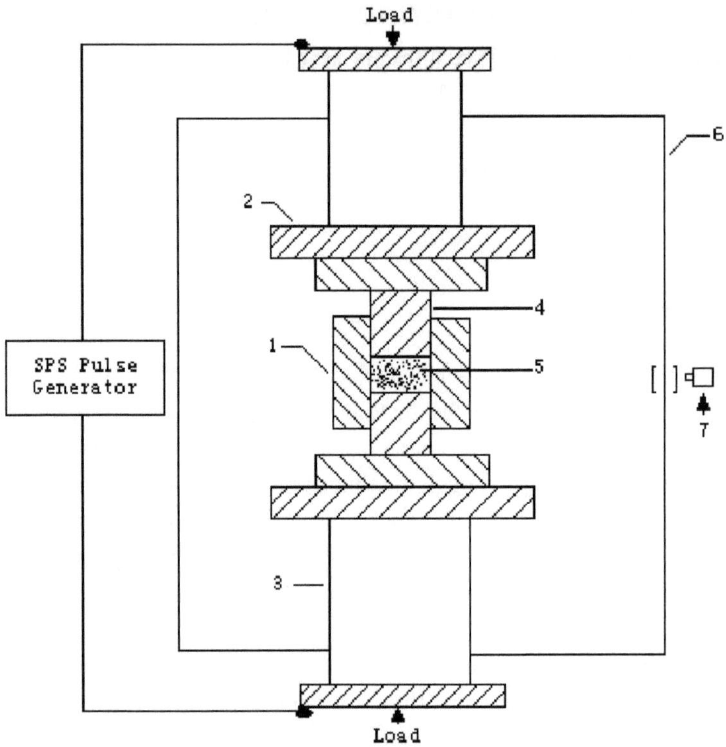

Fig. 8.1. Schematic of a SPS furnace: 1 graphite die, 2 graphite spacer, 3 electrode, 4 punch, 5 sample, 6 vacuum chamber

Fig. 8.2. The spark observed between two TiB_2 particles

For conductive materials, the electrical current runs not only through the punch and die but also through the particles. A spark can be induced between particles by the pulsed current [1,2]. Figure 8.2 demonstrates that a spark can be generated between conductive TiB_2 particles during SPS. However, the densification mechanism for non-conductive materials during SPS is even less well understood. Some investigators suggested that the sintering mechanism for SPS is the same as that for hot pressing [3,4]. They suggested that the powders are heated by the Joule energy produced from the graphite die and punch; nevertheless, this suggestion contradicts the experimental data from our study. For example, the microstructure of the sintered body prepared by SPS is usually not uniform. Furthermore, the heating rate and particle size have a profound effect on the final microstructure. Vapor deposits are often noted after SPS. In the present study, the sintering behavior of the non-conductive particles is investigated. The possible mechanisms during spark plasma sintering are proposed.

8.2 Experimental Work

8.2.1 The Sintering Resistance

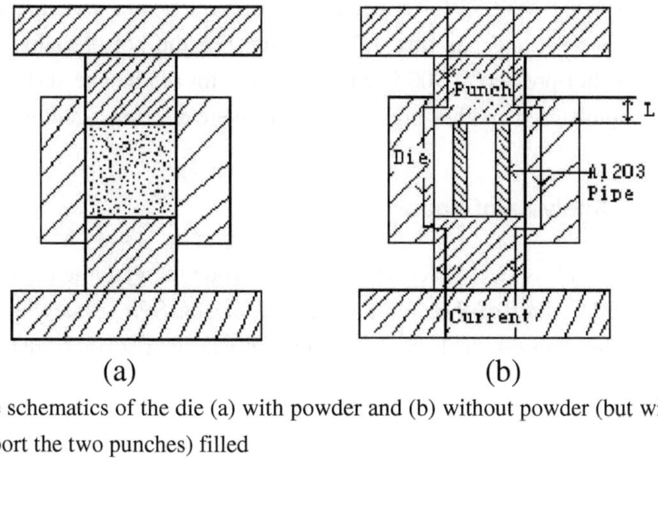

Fig. 8.3. The schematics of the die (a) with powder and (b) without powder (but with Al_2O_3 pipes to support the two punches) filled

SPS was carried out on the graphite die with or without ceramic powders and Fig. 8.3. For the sintering without ceramic powders, the graphite die was heated to 700°C for 1 min with a heating rate of 257°C/min. The temperature was measured by using a thermocouple inserted into the graphite die. By recording the current, I, and voltage, V, during SPS every 30 seconds, the sintering resistance, R, can be calculated as: R=V/I. The sintering resistance of the graphite die only and the graphite die together with Al_2O_3 or BN during SPS is then determined. The heating rate and particle size of Al_2O_3 powder were varied to determine their influence.

8.2.2 The Microstructure

γ-AlOOH powder was used as the raw material in the present study. The powder underwent a series of phase transformations during elevated temperature treatment. The phase evolution of the γ-AlOOH powder can thus be used as the index for the temperature distribution within the sintered body. The γ-AlOOH powder was heated to 580°C, 800°C, 900°C and 1200°C for 0.5 h. The phases were analyzed by XRD. The pressure was 30 Mpa and heating rate 257°C/min. The phases at different depths from the top surface were also analyzed by XRD. The following depths from the top surface were taken, a: top surface, b: 0.3 mm to top surface, c: 1.5 mm to top surface, d: 1.5 mm to bottom surface, e: 0.3 mm to bottom surface, f: bottom surface. Two comparison tests, (1) microwave plasma bombardment test and (2) hot pressing sintering, were also conducted. For the microwave plasma bombardment test, a sample of the same size as that for SPS was formed at 30 Mpa then bombarded by hydrogen plasma at 850°C—900°C for 20 min in a special MWPCVD apparatus. The input power was 1.8 KW, the gas pressure 4.8 KPa, and gas flux 1.2 liter/min. For the hot pressing sintering, the sample was hot-pressed at 900°C under 30 Mpa for 0.5 h. The surfaces of the microwave-bombarded and hot-pressed samples were also analyzed by XRD.

8.2.3 The Analysis of Vapor

A piece of stainless steel sheet was put just outside the graphite die during SPS of Al_2O_3 at 900°C. Part of the vapor generated during SPS was collected from the surface of the stainless steel sheet. The deposit was then analyzed using XPS.

8.3 Results

8.3.1 The Sintering Resistance

If the sintering mechanism of SPS is the same as that of hot pressing, then the sintering resistance should have no reaction to heating rates and particle size. Nevertheless, it was found that: (1) the sintering resistance of the non-conductive powders is higher than that of the empty die; (2) the sintering resistance is affected by the heating rates and particle size.

The sintering resistance of the graphite die with or without non-conductive powders is shown in Fig. 8.4. When non-conductive particles were filled into the die, Fig. 8.3a, the sintering resistance was higher than that of the sintering of the empty die during SPS, Fig. 8.3b. It indicates that the contact resistance between punch and die is increased during the sintering of non-conductive particles. Further investigations show that sintering resistance increases with increasing heating rates. The resistance also increases with decreasing particle sizes (as shown in Fig. 8.5). These results suggest that some interactions took place within the non-conducting powder compacts. As will be demonstrated later, this can be related to the generation of vapor during SPS.

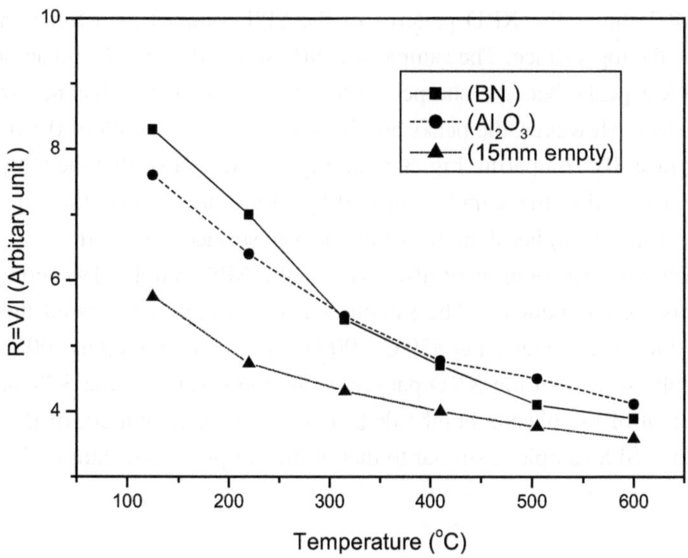

Fig. 8.4. Sintering resistance of the Al_2O_3 and BN compacts and the empty die

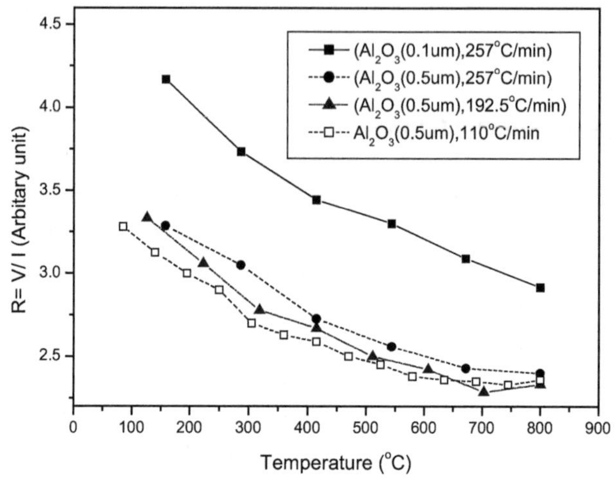

Fig. 8.5. The effect of heating rates on the sintering resistance of Al_2O_3 compacts

8.3.2 The Microstructure

Figure 8.6 shows the XRD patterns of the γ-AlOOH powder compacts heated in a conventional furnace at different temperatures. The phases vary with the increase of temperature.

Figure 8.7 shows the XRD patterns of the SPS compact as a function of the depth from the top surface. The sample was SPS sintered at 900°C. It can be found that the XRD peaks become sharper with the increase of the distance from the bottom surface. However, the peaks are the strongest at the depth of 0.3 mm from the top surface. By comparing Fig. 8.6 and Fig. 8.7, it is clear that the temperature gradient exists within the sample prepared by SPS. Furthermore, the temperature of the top surface is higher than that of the bottom surface during SPS.

Some crystal distortions may also exist in the SPS sample. Two comparison experiments were conducted. The sample was either (1) cold pressed then bombarded by microwave plasma at 850°C—900°C or (2) hot pressed at 900°C for 0.5 h. The results show that the XRD pattern of the top surface of the SPS sample is similar to that of the plasma bombarded sample. The XRD pattern of the bottom surface of the SPS sample is similar to that of the hot-pressesed sample (Fig. 8.8).

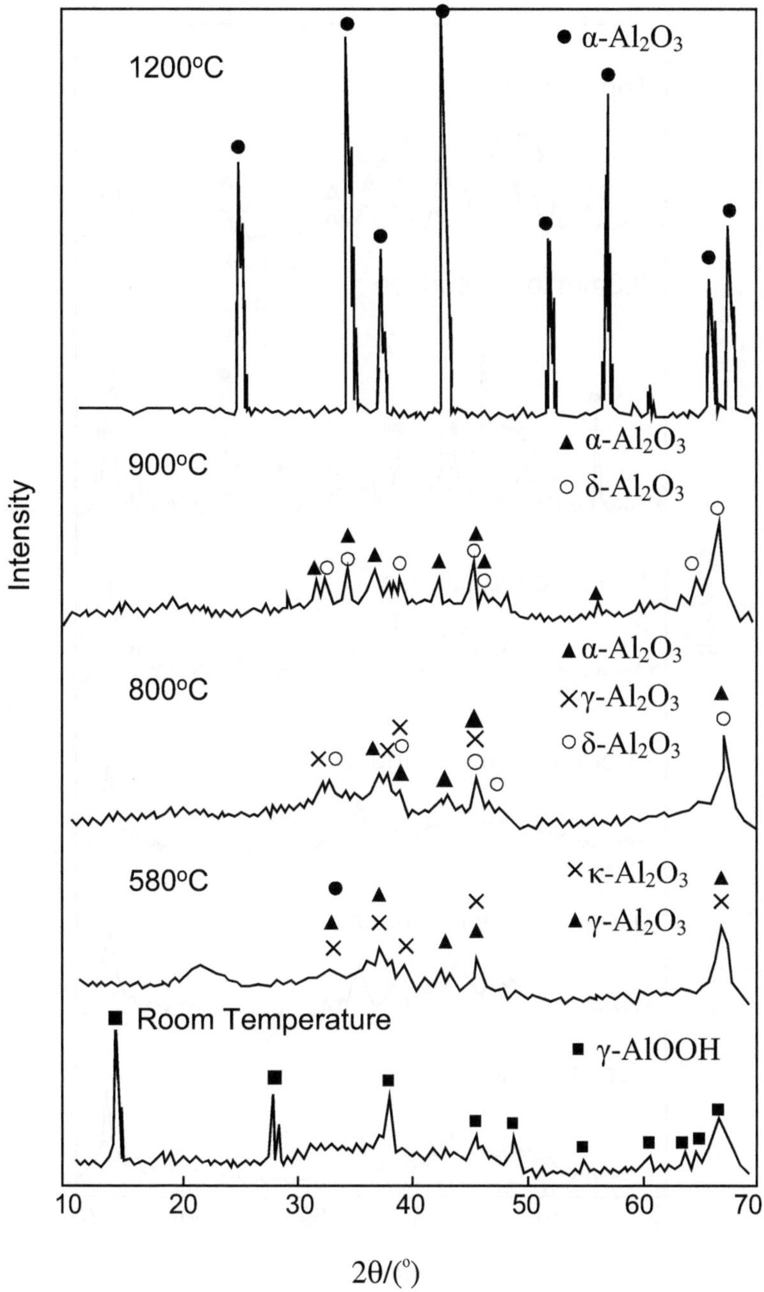

Fig. 8.6. XRD patterns of γ-AlOOH powder heated at different temperatures

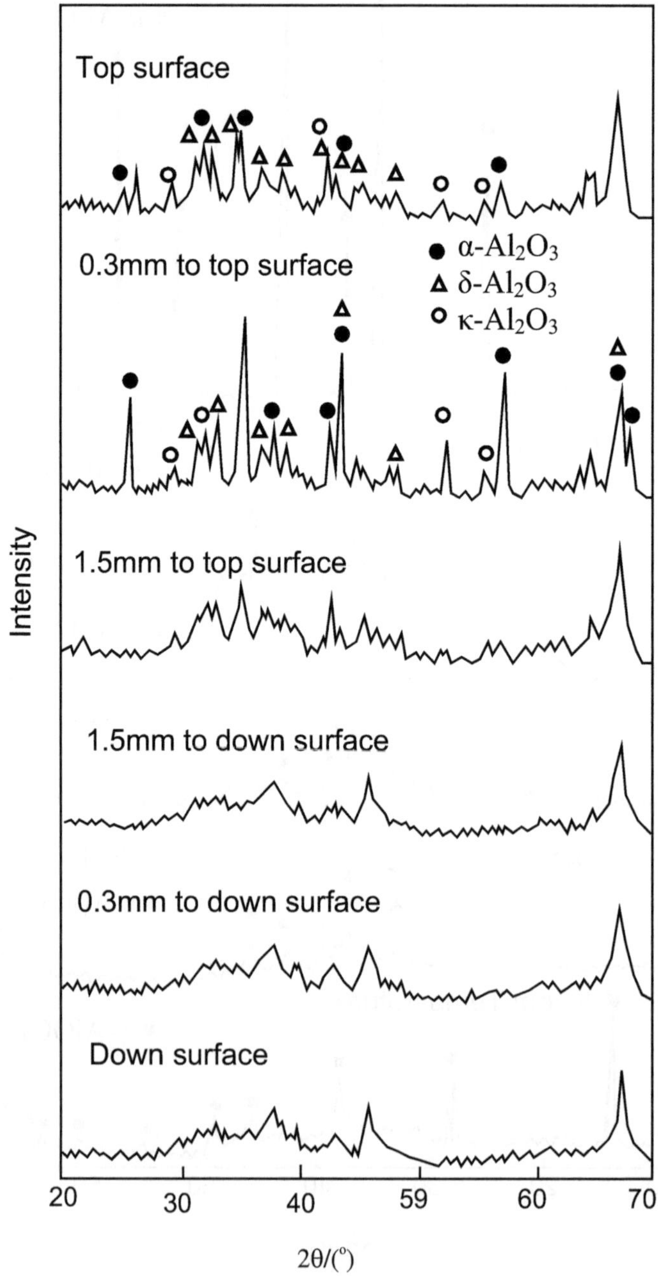

Fig. 8.7. XRD patterns of the sections taken from different depths from the top surface of the SPS Al_2O_3 sample. The SPS was carried out at 900°C

Table 8.1. The Al_{2P} XPS spectrum for the deposit

No	Peak	B.E[eV]	FWHM	G/L	at%	
1	Al_{2P}	72.08	1.5	0.3	8.09	Al
2	Al_{2P}	73.75	1.5	0.3	12.56	$Al_2O_4^{2-}$
3	Al_{2P}	74.64	1.5	0.3	30.08	Al_2O_3
4	Al_{2P}	75.60	1.5	0.3	13.07	AlO_x
5	Al_{2P}	76.60	1.5	0.3	18.67	Al^{3+}
6	Al_{2P}	77.75	1.5	0.3	14.52	

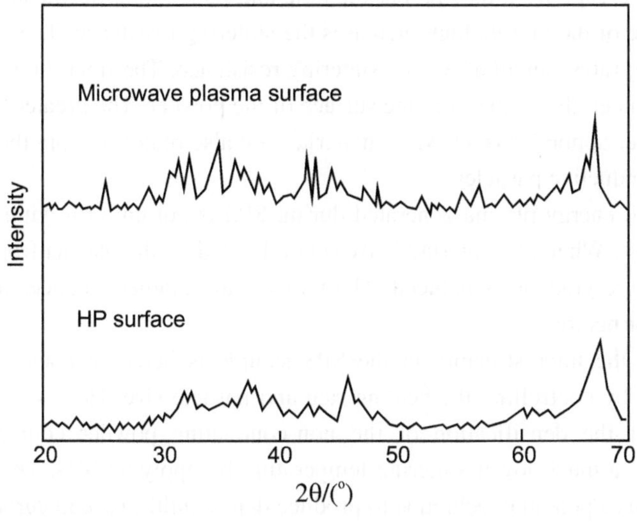

Fig. 8.8. XRD patterns of the Al_2O_3 sample after microwave plasma bombardment and after hot pressing at 900°C

It suggests that the top surface of SPS sample is bombarded by some high-energy particles during SPS. The distortion of Al_2O_3 crystals therefore takes place.

8.3.3 The Analysis on Vapor Deposit

Vapor was usually generated during SPS of an Al_2O_3 sample. Part of the vapor deposit was collected from the surface of a surrounding stainless steel sheet. The XPS results are shown in Table 8.1.

Further investigation into the chemical states of Al by Al_{2P} XPS spectrum was carried out. It was found that more than one chemical state of Al exists in the deposit, which include Al (BE at 72.08 eV), $Al_2O_4^{2-}$, (BE at 73.75eV), Al_2O_3(BE at 74.65eV), AlO_x (BE at 75.6eV) and Al^{3+}(BE at 76.6eV). This demonstrates that Al-oxide particles are formed due to the formation of the high energy plasma. The surface of the particles has also been observed to be more smooth after SPS.

8.4 Discussion

The experimental results suggest that high-energy particles are formed during pulse electric current sintering. Powders are bombarded by the particles and some vapor may escape from the gap between die and punch into the vacuum chamber. The change of the morphology increases the sintering resistance. The particle size and heating rates can all affect the sintering resistance. The finer the particle size, the more materials escape from the surface of the powder. The greater heating rate needs higher applied current. More materials are also removed from the surface of the non-conductive particles.

The high-energy plasma generated during SPS is not uniform within the powder compact. When the sintering body is bombarded by the plasma from one side, a temperature gradient is induced. The temperature gradient induces heterogeneous microstructure.

Though the microstructure in the SPS sample is heterogeneous; this can be minimized by controlling the heating rate and particle size. However, it is worth noting that the densification of the non-conducting powder compact can be achieved at a much lower sintering temperature by applying SPS. The SPS technique is thus a potential technique to produce dense multi-phase materials.

8.5 Conclusions

Two conclusions can be drawn from the present study.

1. High-energy plasma is generated during pulse electric current sintering of non-conductive particles.
2. The formation of high-energy plasma is not uniform within the powder compact; temperature gradient exists. Uniform microstructure can only prepared by carefully controlling the particle size and heating rate.

Acknowledgments

The present study is supported by the National Nature Science Foundation of China (59872024, 59925207) and by the Key Teacher Fund of National Education Ministry.

References

1. Tokita, M. (1999) Materials Science Forum, **308**, .83
2. Kim, H., Kawahara, M., Tokita, M. (2000) J. Japan Soc. of Powder and Powder Metal., **47**, 887
3. Wang, S.W., Chen, L.D., Hirai, T. (1999) J. Materials Science Letters, **18**, 1119
4. Wang, S.W., Chen, L.D., Hirai, T. (2000) J. Mater. Res., **15**, 982
5. Kim, H., Kawahara, M., Tokita, M. (2000) J. Japan Soc. of Powder and Powder Metal., **47**, 887
6. Kaga, H., Carrillo-Heian, E.M., Munir, Z.A. (2000) J. Japan Soc. of Powder and Powder Metal., **47**, 909
7. Gao, L., Wang, H.Z., Hong, J.S. (1999) J. European Ceramic Society, **19**, 609
8. Ozaki, K., Kobayashi, K., Nishio, T. (2000) J. Japan Soc. of Powder and Powder Metal., **47**, 293
9. Nose, M., Nagae, T., Yokota, M. (1998) J. Japan Soc. of Powder and Powder Metal., **45**, 670

9 Pulsed Chemical Vapor Infiltration : Carbon/SiC Nanocomposites

N.-H. Tai, C.-F. Chen, C.-Y. Lee

9.1 Introduction

Over the past two decades, many processing routes for the fabrication of ceramic matrix composites (CMCs) have been proposed [1,2]. Chemical vapor infiltration (CVI) is one of the novel processes developed recently. The process has the potential to produce CMC with superb high temperature performance [2]. For isothermal CVI (ICVI), a fibrous preform is placed into a chamber, the reactants penetrate into the preform and deposit the reaction product(s) onto the fiber surface for forming matrix of the composites. In this process, a deposition gradient due to the concentration gradient is inevitable. To alleviate the drawback, low temperature is preferred. However, a lower processing temperature results in a longer process period.

The forced-flow, temperature-gradient chemical vapor infiltration process (FCVI) can effectively reduce the disadvantage of the extremely long processing period in the ICVI process. In the process, the reactants are forced to flow into the preform by conducting a pressure gradient between the inlet and the outlet. The concentration gradient can be reduced due to the forced flow. Experimental results indicated that the process period can be reduced from a month in the ICVI process to one day in the FCVI process. However, the drawback is that one batch can produce only one component. Moreover, an irregular shape is not preferred in the process.

The pulse chemical vapor infiltration process (PCVI) can reduce the disadvantages of both the ICVI and the FCVI processes. In the PCVI process, the reactant vapors are channeled into the reactor; after a short period, the reacted vapor and remaining reactant are evacuated. The periodic filling and evacuation is repeated throughout the whole process. During the process, the reactants within the preform are refreshed; therefore, the vapor concentration gradient within the preform can be alleviated, which produces ceramic composites with a low matrix gradient distribution. Introducing nanopowder into the ceramic material not only increases the toughness but also alters the fracture manner from brittle failure to plastic deformation[3—5]. Scientists found that the ceramic nanocomposite materials (nanoparticle reinforced) can elongate to up to ten times their original length at elevated temperatures [6]. It is proposed that grain boundary sliding may play an important role in the deformation of ceramic composites. To the knowledge of the present authors, little study deals with the performance of the composites with the incorporation of nanoparticle and continuous fibers as the reinforcement. In this study, SiC nanoparticles and Si particles were added into the precursor of the ceramic matrix; therefore, the influences of the SiC nanoparticles on the continuous carbon fiber reinforced SiC matrix can be studied; and the effects of SiC nanopowder additive on the variation of microstructure of the composites can be examined. A two-step process, involving slurry impregnation and PCVI, was adopted in this work. The first step is the impregnation of matrix precursor into the carbon cloth fabric; the second is the densification of the nanocomposites.

9.2 Experimental Work

9.2.1 Preparation of Specimen

Silicon powder (99.0%, Aldrich) and SiC nanopowder (20 nm, MTI) were introduced to a mixture of PEI (Polyethyleneimine, Alfa Aesar, 99%) and IPA (Isopropyl Alcohol, TEDIA, 99%) which was used as a dispersant. Sonication was used to disperse Si and SiC powders before the slurry was added to the phenolic resin. Eight layers of carbon fiber cloths (W3121, 8 harness satin, Toho Rayon Co., Ltd, Japan) impregnated with the slurry were stacked in a mold. Hot pressing and stabilization were performed at 170°C and 230°C respectively, both under a pressure of 13.8 MPa. Carbonization at 1100°C in a tube furnace was used to evolve the non-carbon species. SiC matrix, which was converted from the reaction sintering of carbonized carbon and silicon, was formed during the heat treatment at 1450°C in a graphitization furnace for 3 hours. In order to enhance the density of the composites, multiple impregnations were applied, heat treatment at 1450°C followed each impregnation.

9.2.2 PCVI Process

To further densify the specimens, PCVI was applied. In the PCVI process, the vapor phase precursor (Tetramethylsilane, TMS, 99.9%, Acros) mixed with carrier gas (hydrogen) was channeled into the reactor. As mentioned, due to the periodic filling and evacuation process, forced flow instead of diffusion dominated the supply of the reactant during the deposition process. The matrix deposition in the PCVI process was affected primarily by the chamber temperature, pressure, precursor concentration, pulsed number, and holding time. In this work, a reaction temperature of 1100°C and a working pressure of 2.5 torrs (holding pressure) were used. The schematic figure of the set-up can be found our published paper in [7].

9.2.3 Mechanical and Physical Properties

Flexural strength was measured by a three-point bending method with a span-to-depth ratio of 20.0 and a crosshead rate of 1.0 mm/min (ASTM D 790). Interlaminate shear strength (ILSS) was measured with the same procedure except that the span-to-depth ratio of 5.0 was used (ASTM D2344). Density and porosity were measured according to ASTM D792 and ASTM C20 standards, respectively.

9.2.4 Microstructure Examination

A field-emission gun scanning electron microscope (FEGSEM, JEOL, JSM-6330F, Japan) was used for the examination of morphologies on both the surface and the fracture surface of the composites. An energy dispersive X-ray spectrometer (EDX) was used for element analysis. Quantitative measurement of carbon and silicon contents in the fabricated nanofiber was carried out. An X-ray diffractometer (D-Max IIB, Rigaku, Japan) was adopted for phase identification. A transmission electron microscope (TEM, JEOL JEM-400EX) was used to examine the crystalline structure.

9.3 Results and Discussion

9.3.1 Mechanical Properties

Interlaminar shear strength (ILSS) and flexural strength of the composites fabricated after each densified step were measured; the results are shown in Figure 9.1. Unfortunately, both ILSS and flexural strength reduce with the increase of impregnation number. From the X-ray diffraction patterns at different depths, it is evident that Si can be detected on the surface and at the sub-layer 1 (i.e. the interface between layers 1 and 2); however, no peak can be detected at the further-deeper layer [7]. We conclude that non-stoichiometry composition occurred on the impregnated slurry, probably due to the different permeability of Si powder and phenolic resin on the open pore in matrix. The non-reacted silicon generates internal thermal stress due to volume change from liquid phase to solid phase during the cooling process. As a result, the degradation of composite after multiple impregnations seems inevitable. However, the mechanical properties can be enhanced after densification through PCVI. Figure 9.2 depicts ILSS of the specimens fabricated under various pulse numbers and reactant concentration at the holding period of 4 seconds. As expected, ILSS increases with pulse number and reactant concentration. This is due to the increase of the density with pulse number and vapor concentration, which is the primary reason for the enhancement of mechanical properties. Monotonic increase in ILSS with pulse number implies that ILSS can be further enhanced by increasing the pulse number and the reactant concentration.

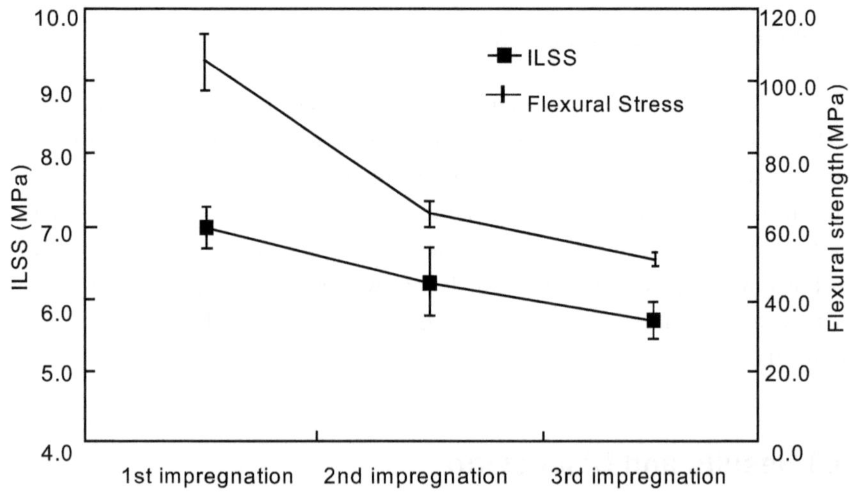

Fig. 9.1. Interlaminate shear strength (ILSS) and flexural strength variation after multiple impregnations

9.3.2 Formation of Nanofibers

The symbol of mSnClK denotes the specimen was fabricated for m seconds at the condition of n% reactant concentration under pulse number of l times 1000 (K) cycles. Figure 9.3 shows the surface morphology of the specimen 2S5C5K. It is detected that sub-micrometer fibers grow from the surface and bridge the particles to form a cross-link structure similar to fibrous mat above the surface. Growth of nanofiber is a heterogeneous reaction, and the bridge process can be performed within the initial several hundred cycles. After examining hundreds of FEGSEM pictures, only a few fiber ends can be observed, most of the nanofibers entangle to form a 3-D structure.

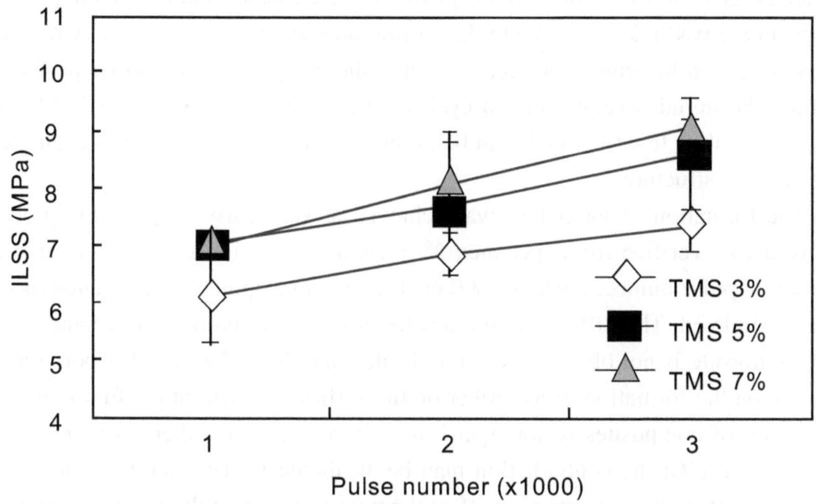

Fig. 9.2. ILSS of the specimens fabricated under various pulse numbers and reactant concentration at the holding time of 4 s

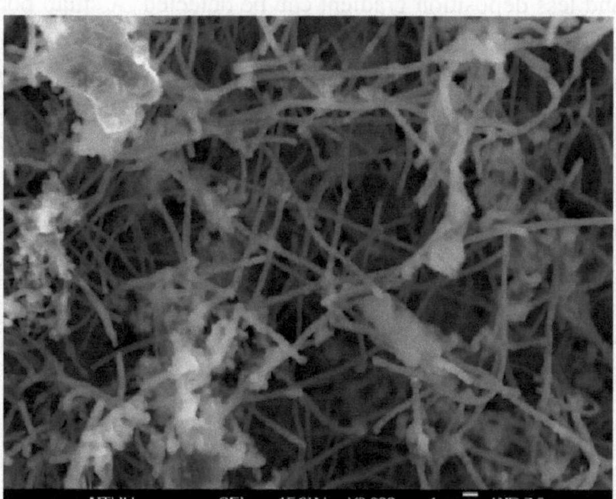

Fig. 9.3. Surface morphology of the specimen 2S5C5K

9.3.2 Formation of Nanofibers

The symbol of mSnClK denotes the specimen was fabricated for m seconds at the condition of n% reactant concentration under pulse number of l times 1000 (K) cycles. Figure 9.3 shows the surface morphology of the specimen 2S5C5K. It is detected that sub-micrometer fibers grow from the surface and bridge the particles to form a cross-link structure similar to fibrous mat above the surface. Growth of nanofiber is a heterogeneous reaction, and the bridge process can be performed within the initial several hundred cycles. After examining hundreds of FEGSEM pictures, only a few fiber ends can be observed, most of the nanofibers entangle to form a 3-D structure.

The formation of nanotubes was achieved at very early stage of the process. This can be verified from specimen 2S5C1K as shown in Fig. 9.4. Even under a very low pulse number such as 1000 cycles, the development of the nanostructure was completed. The difference in diameter of nanofiber on the surface and within the composite is not obvious, which indicates that the influence of vapor concentration on the formation of nanofiber on the surface and within the first interlaminar layer of composites is not significant. However, some other specimens show opposite results; the contradiction may be attributed to the microstructure of the specimen. Pore size and opening affect the transportation of the reactant within the pore network. In the place being examined, if the opening of the pore is large enough, reactants can flow in without obstacle; thus the concentration gradient can be reduced and less deposition gradient can be detected. A small pore opening increases the vapor concentration, which results in a significant deposition gradient. Moreover, nanofiber on the specimen surface shows a rougher surface than that within the specimen. The small node of the nanofiber on the specimen surface is probably due to the inhomogeneous SiC nanopowder additives. When carefully examining the figure, tiny nanofibers can be observed within the sub-micrometer fiber. It is thus reasonable to conclude that nanofiber formation is one of the mechanisms for the growth of sub-micrometer fiber. The deposition of ceramic matrix on the tiny nanofibers may enlarge the diameter of the fiber and then form the sub-micrometer fiber. Moreover, the tip-growth should be the first step and the most feasible mechanism for nanofiber formation.

Fig. 9.4. Morphology within the composite for the specimen 2S5C1K

Fig. 9.5. Spherical nanoparticle on the surface of the specimen without PCVI deposition

Table 9.1. Compositions of the fabricated nanofibers

	EDS integrated intensity		Calculated contents	
	C [%]	Si [%]	C [%]	Si [%]
Mark 1	5.90	4.10	1.50	8.50
Mark 2	1.30	8.70	9.30	0.70
Mark 3	5.70	4.30	5.10	4.90
Mark 4	3.50	6.50	0.40	9.60

As mentioned, a 3-D cross-link structure was developed during the very early PCVI process, the base-growth suppressed by the as-developed structure. Furthermore, the spherical particles, as shown in Figure 9.5, with the nanometer size were detected on the specimen surface. The nanoparticles were in the dimension roughly between 80 and 200 nm. The nanopowder may be derived from the reaction sintering of Si and carbon (converted from pyrolysis of phenolic resin) on the original surface of SiC nanopowder (20 nm). During the PCVI process, the nanoparticle acts as a catalyst for nanofiber formation. The heterogeneous growth due to catalytical reaction on the tip of the nanofibers was further discussed in our published article [7]. In some cases, fewer nanofibers on the surface than within the composites were observed. This fact contradicts the correlation between nanofiber number and vapor concentration. Thus, it is concluded that the number of nanofibers is dependent on the number of nucleation sites on the surface rather than vapor concentration. Concentration merely shows the influence on the growth rate in the radius direction in this study. However, no conclusion can be derived on the correlation between vapor concentration and growth rate of the nanofiber in the axial direction.

9.3.3 Composition and Structure of the Nanofiber

EDX was adopted for composition examination of the deposited nanofiber. For reference, the MAC-3215 SiC standard, SiC nanopowder, and as-fabricated reaction-sintered SiC were measured. All of them are nearly SiC stoichiometry (within 0.3% deviation) [7]. The stoichiometry composition of nanofiber is different from place to place. Table 9.1 lists the composition of the nanofiber measured at different positions. The corresponding positions are shown in Fig. 9.6. It can be seen in the figure that mark 3 was located in the middle section of the nanofiber; marks 1,2 and 4 were located in the tip or on the base of the nanofibers. In this work, it was found that off-stoichiometry composition becomes significant at the tip of the nanofiber and all of them show carbon-rich composition. The carbon-rich composition is probably attributable to the deposited solid products containing extra carbon. The structure of the as-fabricated nanofiber was examined by TEM. Figure 9.7 depicts the layer structure of the nanofiber. It is very clear that although the nanofiber composition is off-stoichiometry SiC, it still shows the layer structure. Further study is required to verify the details of the microstructure of the fiber, especially on the node where the nanotube usually shows a change in growth direction.

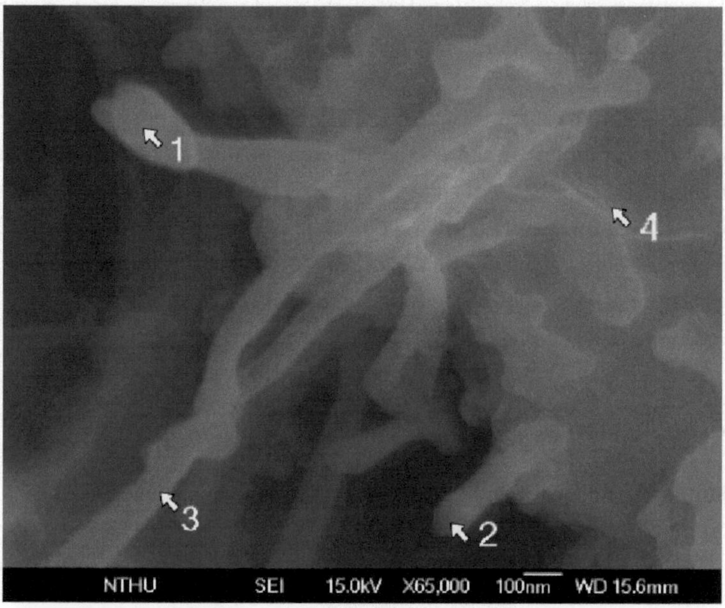

Fig. 9.6. Positions corresponding to the compositions listed in Table 9.1

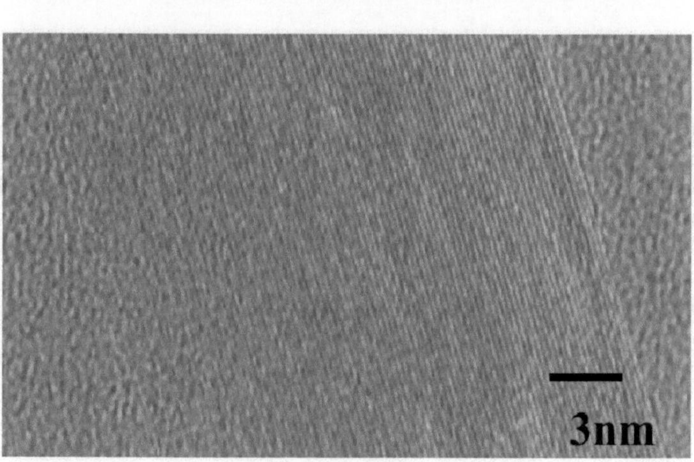

Fig. 9.7. Layer structure of the fabricated nanofiber

9.4 Summary

The mechanical properties of the fabricated ceramic nanocomposites were measured. The mechanical properties decreased with resin impregnation number, due to the generation of internal thermal stress attributed to volume change of the non-reacted Si from liquid phase to solid phase during the cooling process. After PCVI densification, the mechanical property can be enhanced. This is attributed to the enhancement of density with pulse number and vapor concentration, which affects ILSS significantly. Nanofibers and sub-micrometer fibers were formed in the densification process through PCVI. Besides pulse number and vapor concentration, microstructure of the skeleton of the composite also affects the growth of the nanofiber. Based on the structure of the fabricated specimen, a tip growth model is proposed in this work and SiC nanoparticle additives may play a key role for the growth of nanofibers. Off-stoichiometric SiC on the surface of the nanofiber was also detected and the composition changed from place to place. Layer structure of the nanofiber was detected from TEM study.

Acknowledgements

The authors are grateful for the support of the National Science Council, Taiwan under grant No. NSC-89-2216-E-007-031.

References

1. Stinton, D.P., Besmann, T.M., Lowden, R.A. (1988) Ceramic Bulletin, **67**, 350
2. Minet, J., Langlais, F., Naslain, R. (1990) Composites Science and Technology, **37**, 79
3. Wakai, F., Kodama, Y., Sakaguchi, S., Murayama, N., Izaki, K., Niihara, K. (1990) Nature, **344**, 421
4. Ohji, T., Nakahira, A., Hirano, T., Niihara, K. (1994) J. Am. Ceram. Soc., **77**, 3259.
5. Komarneni, S. (1992) J. Mater. Chem., **2**, 1219
6. Wakai, F. (1998) Proceeding of EnCera'98 & HT-CMC3, Osaka, Japan
7. Tai, N.H., Chen, C.F. (2001) J. Am. Ceram. Soc., **84**, 1683

10 Metal Oxidation Process : $NiAl/Al_2O_3$ Composites

C.-S. Hwang, T.-J. Liu, J.-P. Shich

10.1 Introduction

The processing of the ceramic/metal composite with a uniform microstructure is a challenging task. The difficulties mainly stem from the difference in density and particle size between metal and ceramic materials. In the present study, a nano-ceramic powder is introduced into a metallic matrix to enhance its mechanical properties. The processing conditions to prepare a uniform microstructure are explored. The system used to demonstrate the feasibility of the present novel process is a $NiAl-Al_2O_3$ system.

The alumina particles are formed by the oxidation of an Al-containing intermetallic. The nano-sized alumina are formed uniformly at the surface of the intermetallic particles. The distribution of alumina is thus uniform. Nickel aluminide (NiAl) is a potential structural material because of its high melting point (1638°C), high modulus, low density (5.90 kg/m^3), and better oxidation resistance than the nickel-base alloys. Furthermore, the thermal conductivity of NiAl is 3 to 8 times that of Ni-base superalloys [1,2]. However, the polycrystalline NiAl exhibits poor fracture toughness at room temperature and low strength at elevated temperature. To improve these drawbacks, adding Al_2O_3 particles into NiAl matrix to form $NiAl/Al_2O_3$ composites is a promising alternative.

Several preparation methods of $NiAl/Al_2O_3$ composites have been reported, such as powder cloth (PC) technique from the raw materials of stoichiometric NiAl powders and the single-crystal Al_2O_3 fibers [3,4]; attrition milling and grinding of Al_2O_3 and NiAl powders in ethyl alcohol [5]; heat treatment and de-carbonization method from organic precursor $[Ni(N_2H_4)_2](C_6H_5COO)_2$ + $Al(C_6H_5COO)_{1.7}(OH)_{1.3}$ [6]; and mechanical alloying of Ni and Al powders [7,8]. Functional gradient $NiAl/Al_2O_3$ composites have also been fabricated by a modified sedimentation process [9], or by reaction hot compaction of Ni and Al powders in which one or both powders were pre-oxidized [10].

Methods to prepare $NiAl/Al_2O_3$ composites involving sedimentation in water, ball milling and then oxidation of NiAl powder in air to form Al_2O_3 are explored in the present study. Effects of NiAl particle size, ball milling time and oxidation condition on the microstructures and mechanical properties of NiAl/Al_2O_3 composites are investigated.

10.2 Experimental Work

NiAl powders with average particle sizes of 140 μm and 5.9 μm were used to prepare NiAl/Al_2O_3 powders. The larger NiAl powders were first ball-milled for 24 h into ~12 μm. The powder was then oxidized at 700—1000°C for 24 h in air. The small NiAl powder was further classified into 3, 5 and 7 μm particles by a sedimentation process in deionized water at room temperature. After centrifuging and drying at 100°C for 24 h, the powders were oxidized at 600—700°C for 2 h in air. The NiAl/Al_2O_3 powders were then put into the graphite mold, and hot pressed at 1400°C for 1 h under a vacuum of 10^{-2} Pa, with a pressure of 25 MPa.

Hot-pressed specimens were cut and machined longitudinally with a diamond wheel, and ground into 4 mm× 4 mm× 35 mm. The surface was then polished with 1 μm diamond paste and chemically etched. The bulk density was determined by Archimedes' method. Phase identification and lattice constant determination were performed using X-ray diffraction analysis (XRD, Rigaku D/Max-IIB). The microstructures were observed by scanning electron microscopy (SEM, Hitachi S-2500) and transmission electron microscopy (TEM, JEOL JEM-3010). The room temperature strength of the specimens was determined by the four-point bending technique with a cross-head rate of 0.5 mm/min. The hardness was determined by Vickers hardness tester. Fracture toughness was determined by the single-edge notched beam (SENB) technique. An image analyzer was used to determine the contents of Al_2O_3 in the NiAl/Al_2O_3 composites. Specimens are denoted as xNAy, where x stands for the oxidation temperature (× 10°C), and y for the particle size of NiAl powders (NA) .

10.3 Results and Discussion

10.3.1 Ball Milling Followed by Oxidation

TEM micrographs show the morphology of the larger NiAl particles that were oxidized at 700°C for 2 h. The diffraction patterns of the indicated areas in

Fig. 10.1 (a) and (b) are also shown in the figure. A laminated oxidized layer around the particle is observed. The diffraction patterns show that the structure of the as-received NiAl particle is bcc, and the oxidized layer is composed of amorphous and microcrystals. The thickness of the oxidized layer analyzed by AES increases with increasing oxidation temperature. The layer then crystallizes into α-Al_2O_3 at an elevated temperature. Figure 10.2 shows the microstructures of the polished and chemically etched surface of hot-pressed NiAl/Al_2O_3 composites prepared from the NiAl powder oxidized at 700°C and 1000°C for 2 h. The amount of second phase at the grain boundary of NiAl increases with increasing oxidation temperature. From the XRD analysis, only NiAl and α-Al_2O_3 phases are found in the hot-pressed composites, i.e. the second phase is α-Al_2O_3. The α-Al_2O_3 forms a more or less continuous network around the NiAl grains after hot-pressing. The presence of the Al_2O_3 particles inhibits the grain growth of NiAl.

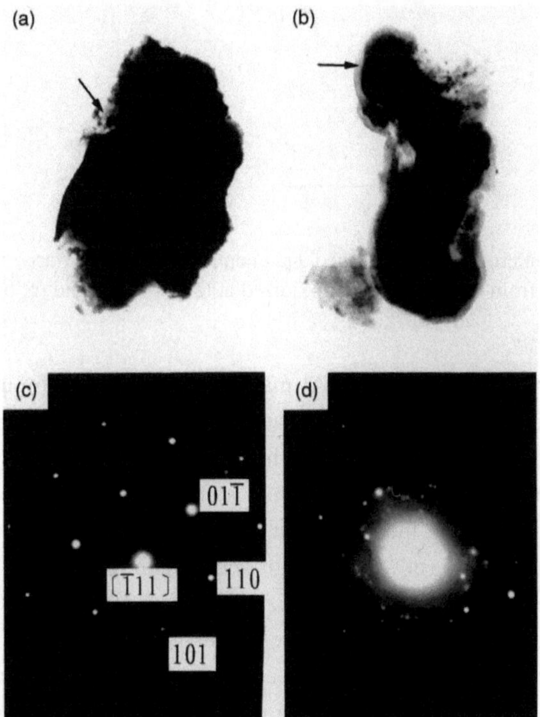

Fig. 10.1. TEM micrographs of (a) the NiAl particles as received and (b) then oxidized at 700°C for 2 h. The electron diffraction patterns from the arrows in (a) and (b) are shown in (c) and (d), respectively

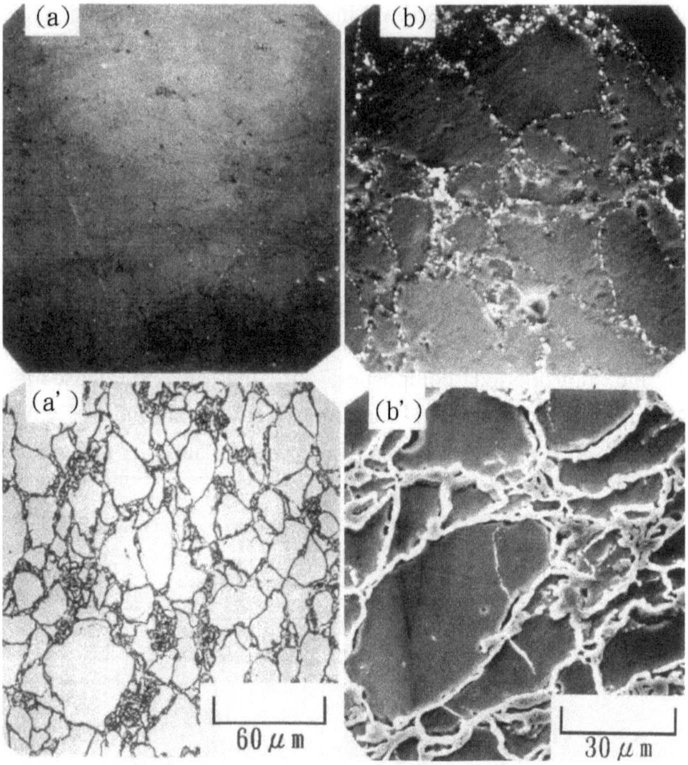

Fig. 10.2. SEM micrographs of polished and chemically etched surface of NiAl/Al_2O_3 composites prepared from the NiAl powder oxidized at (a, a') 700°C and (b, b') 1000°C for 2 h

The Al_2O_3 particles between NiAl matrix grains are also found in the sample prepared by the NiAl powder pre-oxidized at 1000°C. Figures 10.3 and 10.4 show the room-temperature bending strength and fracture toughness of NiAl/Al_2O_3 composites, respectively. For the bending strength and toughness, the specimen created by first oxidizing the powder at 700°C for 4 h has the best values: about 1000 MPa and 15.4 $MPa \cdot m^{1/2}$ for strength and toughness, respectively. The values of the specimens created by first oxidizing the powder at >800°C decrease with increasing oxidation temperature. These results can be explained in terms of the above microstructural characteristics, i.e. Al_2O_3 content, grain size, and shape of the NiAl matrix grains. In the oxidation treatment, Al_2O_3 is formed on the surface of the NiAl particle, and then dispersed at the grain boundary of NiAl/Al_2O_3 composites. The bending strength and toughness change with oxidation temperature can mean that the residual thermal stress between $NiAl_{1-x}$ matrix and Al_2O_3 is so large that the larger defects are found at the grain boundary (Fig. 10.2(b')), which reduces the bending strength and toughness.

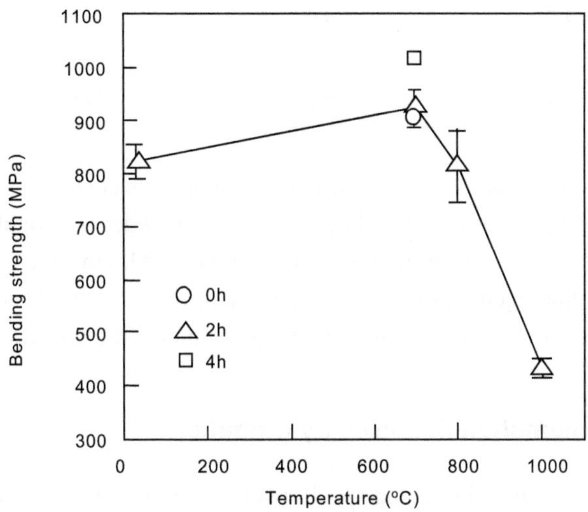

Fig. 10.3. Bending strength of hot-pressed NiAl/Al_2O_3 composites prepared by first oxidizing NiAl powder at the indicated temperatures for various times

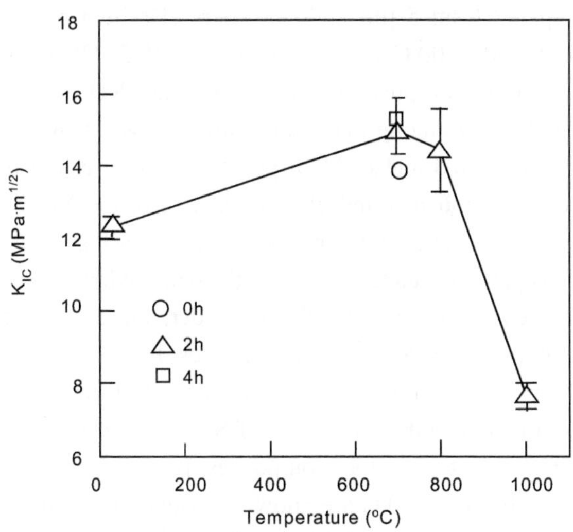

Fig. 10.4. Fracture toughness of hot-pressed NiAl/Al_2O_3 composites prepared by first oxidizing NiAl powder at the indicated temperatures for various times

Table 10.1. The particle size and specific surface area of the classified NiAl powders

Classified	Median diameter [μm]	Modal diameter [μm]	Specific surface area [m^2/g]
3 μm	4.0	4.6	4.88
5 μm	6.2	6.5	4.23
7 μm	8.8	9.8	3.77

From the above data and microstructure, the process consisting of ball milling, oxidizing NiAl powder and then hot-pressing the powder is a viable technique to improve the mechanical properties of monolithic NiAl. However, the NiAl grain sizes are not homogeneous (uniform). To obtain the NiAl/Al_2O_3 composites with homogeneous NiAl grain size, the following sedimentation process was explored.

10.3.2 Sedimentation Followed by Oxidation

The median size, modal size and specific surface area of the 3, 5 and 7 μm classified powders selected from the as-received NiAl powder (5.9 μm) are shown in Table 10.1. The classified and as-received powders are oxidized at 600—700°C in air to form NiAl/Al_2O_3 powder and then hot pressed at 1400°C for 1 h under vacuum. The representative X-ray diffraction patterns of hot-pressed NiAl/Al_2O_3 composites prepared from 5 μm NiAl powders classified by sedimentation and then oxidized at 600—700°C are shown in Fig. 10.5. Only NiAl and α-Al_2O_3 phases are found in the hot-pressed composites. The XRD intensity of the 70NA5 composite is the highest among the composites. The XRD analysis shows that no phase other than the NiAl and α-Al_2O_3 phase formed after hot pressing. Typical SEM micrographs of polished and etched surfaces of the NiAl/Al_2O_3 composites are shown in Fig. 10.6. The white and fine particles in Fig. 10.6 are Al_2O_3, which are discontinuous particles dispersed within the dense NiAl matrix. Apart from the specimens prepared from the as-received powder, the composites show narrow grain size distribution of NiAl and Al_2O_3 particles. Each NiAl grain can be distinguished clearly due to the Al_2O_3 particles' tendency to locate at the grain boundaries. The Al_2O_3 content and the grain sizes of NiAl can be measured from the SEM photographs. The results for the composites prepared from 3, 5 and 7 μm NiAl powders oxidized at 650°C are shown in Table 10.2. The Al_2O_3 content increases slightly with decreasing NiAl particle sizes. This suggests that the extent of oxidation of small NiAl particles is more severe than that of the large particles after the same oxidation treatment.

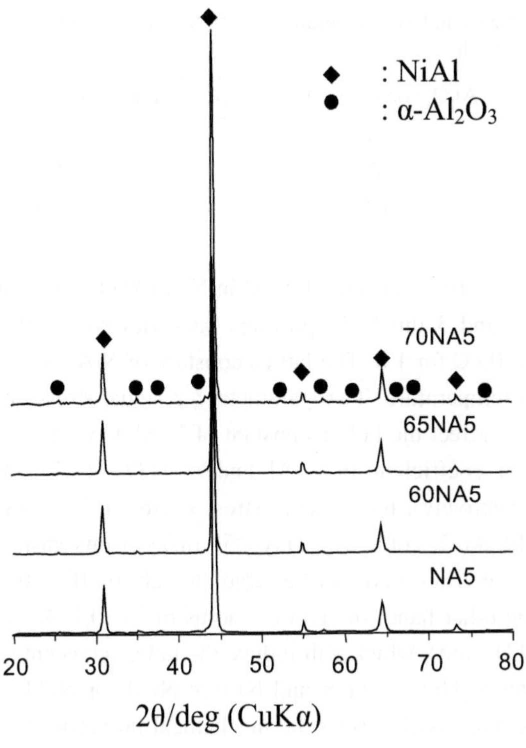

Fig. 10.5. XRD patterns of hot-pressed $NiAl/Al_2O_3$ composites prepared with 5 μm NiAl powders and pre-oxidized at 600—700°C

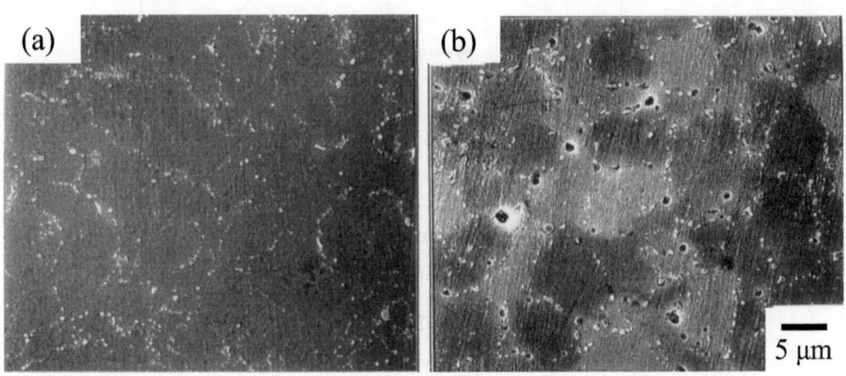

Fig. 10.6. SEM micrgraphs of polished and etched surfaces of $NiAl/Al_2O_3$ composites prepared from the (a) as-received and (b) 7 μm NiAl powders and oxidized at 650°C for 2 h

Table 10.2. Al_2O_3 contents and average grain sizes of NiAl in the NiAl/Al_2O_3 composites hot-pressed at 1400°C for 1 h

Samples	Al_2O_3 content [vol%]	Average NiAl grain size [μm]
65NA3	5.1	7.1
65NA5	4.8	8.4
65NA7	4.4	10.0

Figure 10.7 shows the lattice constant of NiAl in NiAl/Al_2O_3 composites prepared from the as-received and 3 μm NiAl powders pre-oxidized at 600—700°C and then hot pressed at 1400°C for 1 h. The lattice constant of NiAl decreases with the increasing oxidation temperature. Two parameters, residual stress and solid solution, are considered to affect the lattice constant of NiAl. Owing to the difference in thermal expansion coefficient of NiAl and α-Al_2O_3 (12.7—15×10^{-6}/K and 7.2—8.6×10^{-6}/K, respectively), the residual stress exists in the hot-pressed composites. However, the Al_2O_3 content is only <5 vol.% in this case. The residual stress induced from the Al_2O_3 may not be large enough to affect the lattice constant of NiAl. On the other hand, the atomic radius of Ni (0.1246 nm) is smaller than that of Al (0.14315 nm), which will induce the lattice constant of NiAl to reduce as the oxidation of NiAl occurrs and Ni-rich NiAl (or $NiAl_{1-x}$) forms. The change in lattice constant would affect the mechanical properties of the composites.

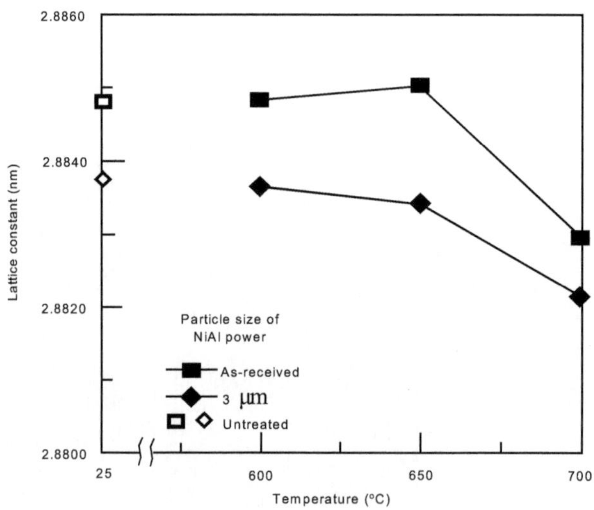

Fig. 10.7. Lattice constant of NiAl in the NiAl/Al_2O_3 composites prepared with the NiAl powder pre-oxidized at 600—700°C and then hot pressed at 1400°C for 1h

Figure 10.8 shows the hardness and room-temperature bending strength of NiAl/Al_2O_3 composites with different NiAl particle sizes and pre-oxidation temperature of NiAl powder, respectively. The hardness increases with increasing oxidation temperature and with decreasing NiAl particle sizes. The composite prepared from 3 μm NiAl powders and then oxidized at 700°C for 2 h shows the highest hardness value of 483 Hv. The increase in hardness is owing to the high Al_2O_3 content. The strength of the composite prepared from the as-received NiAl powder is only 640 MPa. The composites prepared from classified and oxidized NiAl powders increases to a range of 930—1200 MPa. For the specimens prepared from the same particle size, the strength increases slightly with increasing oxidation temperature (i.e. increasing Al_2O_3 contents). At the same oxidation temperature, the strength increases with decreasing particle size of NiAl. These results show that narrow grain size distribution of NiAl and Al_2O_3 particles, and optimum Al_2O_3 contents, are obtained by the sedimentation and then oxidation treatment of NiAl powder, which enhance the mechanical properties of NiAl/Al_2O_3 composites.

10.4 Conclusions

The preparation, microstructure and mechanical properties of the NiAl/Al_2O_3 composites, prepared from the NiAl powder by (1) ball milling and then oxidation and (2) sedimentation and then oxidation were investigated. The following conclusions can be made.

1. The Al_2O_3 content increases with increasing pre-oxidation temperature and with decreasing size of NiAl particles.
2. The Al_2O_3 dispersed at the grain boundaries of NiAl grains.
3. The mechanical properties of NiAl/Al_2O_3 composites were enhanced by optimizing the content and size of Al_2O_3 particles.

Acknowledgment

Financial support of this research by the National Science Council, Republic of China, under contract no. NSC-83-0405-E-006-002, NSC-85-2216-E-006-029, and helpful discussion with Dr. W. H. Tuan and Dr. S.T. Chang, are gratefully acknowledged.

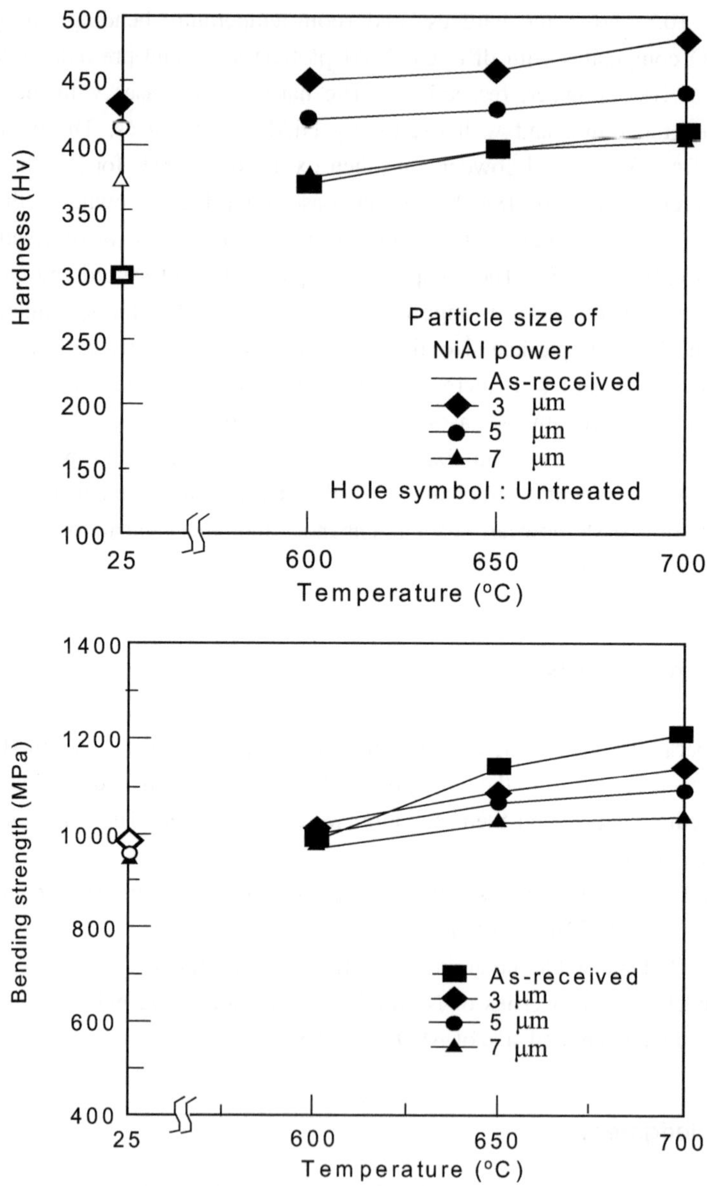

Fig. 10.8. Strength and hardness of $NiAl/Al_2O_3$ composites as a function of the pre-oxidation temperature of NiAl powder

References

1. Noebe, R.D., Bowman, R.R., Nathal, M.V. (1993) Int. Mater. Rev., **38**, 193
2. Darolia, R. (1991) J. of Metals, **43**, 44
3. Pickens, J.W., Noebe, R.D., Watson, G.K., Brindley, P.K., Draper, S.L. (1989) NASA Technical Memorandum No. TM-102060, NASA, Washington, DC, USA
4. Wang, L., Xu, K., Bowman, R.R., Arsenault, R.J. (1997) Metall. Mater. Trans., **28A**, 2755
5. Tuan, W.H., Pai, Y.P. (1999) J. Am. Ceram. Soc., **82**, 1624
6. Abe, O., Takata, S. (1999) Key Eng. Mater., **159/160**, 319
7. Dollar, M., Dymek, S., Hwang, S.J., Nash, P. (1993) Metall. Trans., **24A**, 1993
8. Doi, Y., Akimoto, H., Matsuki, K., Aida, T., Ochiai, S. (2000) J. Japan Inst. Metals, **64**, 518
9. Miller, D.P., Lannutti, J.J., Noebe, R.D. (1993) J. Mater. Res., **8**, 2004
10. Hwang, C.S., Liu, T.J. (1999) J. Mater. Res., **14**, 75

References

1. [illegible]
2. [illegible]
3. [illegible]
4. [illegible]
5. [illegible]
6. [illegible]
7. [illegible]
8. [illegible]

11 Co-electroplating Nano-ceramic Particulates with Ni Layer

S.-C. Wang, W.-C. J. Wei

11.1 Introduction

There has been a tremendous upswing in the progress of nano-technology over the past decade. Massive research subjects about nanostructured materials are proposed and become potentially important for advanced development of many new fields. The objects are constructed of units in the size range from 1 to 100 nm**, including nanoparticles, whiskers, nano photonic crystalline, nanotubes and nano films all of which have been demonstrated useful in the fields of bio-materials, micro-electronics and micro-electro-mechanical system (MEMS) [1].

There are many methods have been documented to manufacture nanostructured materials [2,3]. One approach starts from ultra-fine powders. The powder may go through shape forming and densification. However, one severe problem is reported: that the optimal microstructure is incapable of being retained after inappropriate consolidation or long-term service at high temperature. Sophisticated control and efforts have to be put into the stage of densification of the specimen in order to keep the original merits of nanostructured materials.

For a nanocomposite containing nickel matrix and abrasive ceramic phases, the electrochemistry route is one convenient and economic way to reach full density without high temperature treatment [4]. By controlling the electroplating parameters, it is possible to synthesize nanostructured materials in a pore-free state at ambient temperature. The simultaneous deposition of hard particles and metal phases to form composite coating results in an improvement of physical and mechanical properties of the metal layer. The composite plating including hard particles has potentially for today's industrial applications of precision molds (DVD or CD), electroforming coating in the LIGA process, automotive industries, or the electrical contacts in electronic parts [5].

The added ceramic particles reported [6—13] include Al_2O_3, SiC, Cr_2O_3, TiO_2, MoS_2 and WC, etc. The hard particles in general improve the hardness and tri-

** In this paper, the terms nanosized, nanostructured and ultrafine means the features in the sizes less than 100 nm but larger than atomic scale.

bological properties of the coated layer by the dispersive strengthening mechanism. The hard inclusions inhibit the movement of grain boundaries and slow down the grain growth during thermal treatment. When the material suffers an external applied stress, the inclusions retard the dislocation movement in metallic matrix [14], a plastic deformation becomes difficult so the hardness of the composite layer increases and the wear resistance can be improved. This can be still further improved by reducing the particle size of included ceramic particles [15]. The combined effects are thought to be enhanced by using nanoparticles in the electro-plating composites.

When ultrafine particles are dispersed into the plating solutions of nickel to form composites by the electrochemical route, the stability of nanoparticles in suspension is controlled by the colloidal properties, e.g. the thickness of the double layer and the ζ–potential of the particles. If the suspension shows poor dispersive states, the composites would have agglomerated inclusions, which offer weak points or form stress concentration spots during fracture test. Especially in the high temperature environment, the composites contain agglomerated inclusions show a poorer ability of grain-boundary pinning than that those uniformly dispersive inclusions. Thus, the surface charge of the powders in suspension is an important issue for the mechanical properties of the composites. Another important issue of electrokinetic properties of the suspension is that, during the composite plating, the ultrafine particles and the metal ions must show positive charge, so that the codeposition of metal and inclusions onto the cathode becomes possible. The colloidal property of SiC particles in the Ni plating solution and the wear resistance of the SiC composite layer have been reported previously [16—18]. In this study, we tried to classify the suspension and adsorption steps of the SiC particle in the Ni plating solution and proposed a composite plating mechanism. Then, we used ultrafine SiC, Al_2O_3 or ZrO_2 powders to codeposite with Ni to form composites. The microstructural evolution of the Ni matrix as the ultrafine ceramic particles added was characterized by analytical TEM. The goals focused on the grain size distribution, the preferred orientation and the interface of the Ni to ceramic particles.

11.2 Experimental Procedures

The ultrafine ceramic powders used in this study included SiC, Al_2O_3 and ZrO_2. The isoelectric points (IEP), the mean particle size, and phases of the three kinds of the powders are listed in Table 11.1.

Table 11.1. Properties of the three powders used in this study

Property	SiC (PJ-PL-SiC)	Al_2O_3 (PJ-Pl-AlO)	ZrO_2 (PJ-PL-ZrO)
IEP	3	9	7.5
Average agglomerated size [μm]	0.22	0.08	0.06
Maximum agglomerated size [μm]	0.5	0.5	0.2
Crystalline size (by TEM) [nm]	20 – 200	20-80	5-25
Crystalline phases	6H phase	δ phase	Cubic
Dispersibility in the solution of pH=4	Difficult	Easy	Medium

The composite plating solution used a nickel sulfamate bath containing 90 g/L nickel sulfamate (EP grade, Phibrochem Inc., USA), 3 g/L nickel chloride (EP grade, SHOWA, Japan), and 40 g/L boric acid (EP grade, SHOWA, Japan). The composites were plated on a Cu cathode with dimensions of 10 cm× 6.5 cm× 0.25 cm. The anode was a Ti basket containing small Ni ingots. Before Ni plating, the Cu plate was always ground and etched by diluted sulfuric acid to remove the oxides and grease on the surface.

The amount of the powder added into the plating solution was 1 vol.% based on the bath volume. Gentle stirring of the suspension by air bubbles kept the solution well dispersed. The current density was 4 A/dm^2 (ASD). The bath temperature was controlled at 50±2°C and the pH ranged from 4.0 to 5.0, which was adjusted by H3BO3 or NiCO3. The surfactant of sodium hexanitrocobaltate $[Na_3Co(NO_2)_6]$* (Jassen Chimica, Beerse, Belgium) was used in the composite plating of the SiC/Ni system to increase the codeposition rate of SiC [20].

Crystalline phases of the composite layers were identified by X-ray diffractometry (PW1792, Philips Instrument, Netherlands). The codeposition layer was cut and polished, then the cross-section was examined with scanning electron microscopy (SEM, XL30 Philips, Netherlands) equipped with EDS (EDS DX4, EDAX Corp., USA). A thin foil sample of 3 mm diameter prepared from the coating layer was prepared and characterized by transmission electron microscopy (TEM, 100CXII, or HRTEM, FX-400E, JEOL, Japan).

*Patented number 87110417, ROC.

11.3 Results and Discussion

11.3.1 Modeling of Composite Plating

While the nanosized particles are suspended in a plating solution, the surface of the ceramic particles react with H^+, OH^- ions or available species in solution, and form a charged surface. Divalent cations Ni^{+2} in the plating suspension may adsorb on a negative surface of SiC in a weak acidic solution, and reverse the particle polarity to positive [20]. Some particles will form clusters during the electrophoresis process to the cathode due to the causes of potential instability. When the particles reach the outer Helmholtz plane (OHP) of the cathode driven by convection or electrophoresis, they may overcome a weak energy barrier to form a loose adsorption layer. The phenomenon can be described by the Langmuir adsorption theory [6] and an electrochemical reaction, by which parts of the Ni^{+2} ions adsorbed on the surface of ceramic particles are reduced to Ni.

We have reported that the deposition of charged ultrafine SiC particles on the cathode by the assistance of $[Co(NO_2)_6]^{-3}$ complex ions is the major controlling step for particle codeposition [20]. An easier charge transfer during the transformation from loose to strong codeposition [6] makes SiC adsorbed on the cathode at higher concentration. Then the SiC in clusters are engulfed by reduced Ni and form a consistent plating layer.

A model of the codeposition mechanism of nanosized SiC in the $Ni(SO_3NH_2)_2$ solution is proposed, and shown in Fig. 11.1. The codeposition procedure can be classified into five steps: hydration of particles, adsorption of Ni^{+2} on particles, agglomeration of ultrafine SiC particles, diffusion of the SiC toward the cathode, and reduction and codeposition of SiC with Ni to form a composite layer.

Hydration: There is always, for instance, a thin SiO_2 layer on the surface of SiC particles. The hydrophilic SiO_2 surface would adsorb protons to form a positive-charged surface or hydroxyl ions to a negative-charged surface, depending on the pH value of the suspension [19]. In the plating solution of this study, the pH value, 5, is larger than the IEP of SiC, 3. Thus, the SiC would process a reaction shown as follows:

$$\text{Si-OH} + OH^- \rightarrow \text{Si}-O^- + H_2O \quad (11.1)$$

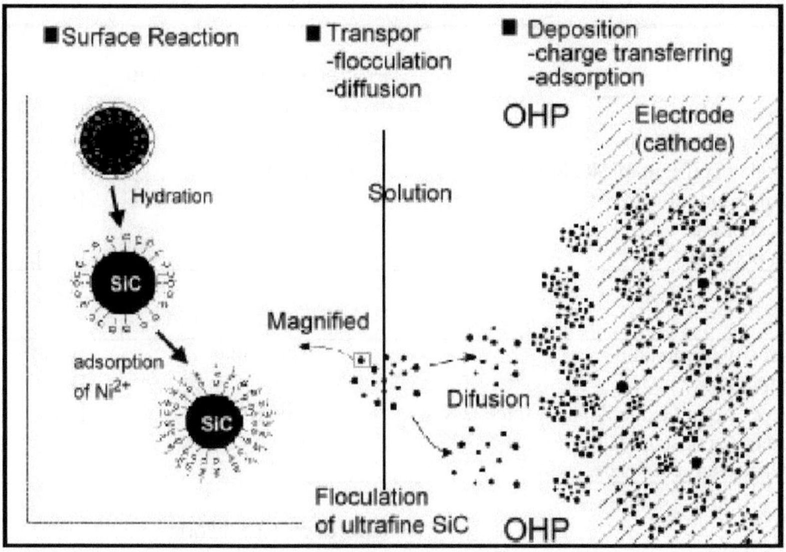

Fig. 11.1. Schematic diagram of a model of composite plating of ultrafine SiC particles in $Ni(SO_3NH_2)_2$ plating solution

From the rate constant of (11.1), pK~6.87 [20], the de-protonation reaction occurred in a suspension of pH=5. Therefore,

$$\log\left[\frac{[SiO^-]}{[SiOH]}\right] = pK_2 - pH = 1.87 \qquad (11.2)$$

the concentration of $[SiO^-]$ on the surface of SiC is about 74 times higher than [SiOH]. The reaction results in a negatively charged surface of SiC particles.

Adsorption of cations: From our previous investigation [18] conducted in a plating solution, one possible adsorption reaction is that the negative surface of SiC will adsorb the cations from the solution and reverse the electrokinetic potential from negative to positive. The reaction can be shown as:

$$Si\text{-}O^- + Ni^{2+} \rightarrow Si\text{-}O^-\text{-}Ni^{2+} \qquad (11.3)$$

Agglomeration: Microstructure evidence shows that SiC particles in the plating layer appear to have some degree of agglomeration [20]. The double-layer thickness of SiC particle is compressed by very high electrolyte concentration in the plating solution and results in a lower ζ-potential. In order to resolve the effects of surface potential, a discrete element modeling (DEM) simulation was conducted. The calculation conditions are specified in Table 11.3. Particle concentration in a

Table 11.2. Physical and chemical properties used for DEM simulation

	Properties of SiC Particle	
Density of SiC	3.2×10^3	[kg/m³]
Density of medium	1.0×10^3	[kg/m³]
Hamaker constant of SiC in water	10.90×10^{-20}	[J]
Young's modulus of SiC	4.0×10^{11}	[GPa]
Poisson ratio[4]	0.16	
Zeta potential	0 - 30	[mV]
	Properties of Medium	
Viscosity of medium	1.0×10^{-3}	[Pa.s]
Temperature	300	[K]
Debye length (1/double layer thickness, κ)	1.0×10^8	[m]
Dielectric constant of medium at 25°C	80.1	
Electric charge valence	2	
	Operation Conditions	
Particle concentration	5	[vol.%]
Particle size distribution*	30~300	[nm]
Outer electric field strength	56.25	[V/m]

* The distribution of particles is measured by particle sizer (Malvern 2000, England).

number of 1000 was chosen to reduce the simulation time of one set of data to a reasonable period of 24 h.

The images of DEM simulation for 0 mV or 30 mV of ζ-potential are shown in Fig. 11.2. The degree of agglomeration in volume percentage is estimated from the simulated images in Fig. 11.3. The results indicate that part of the separate particles, especially for the larger particles, are likely to form a cluster structure. This phenomenon contradicts the general concept that finer particles tend to agglomerate more easily than coarser ones. This implies that a Brownian motion may not take into effective. If a suspension shows the conditions described by Table 2 [21,22], the average translation kinetic energy of a particle for one dimension is $0.5kT = 2.2 \times 10^{-21}$ *J*, which is much greater than the kinetic energy for charge effect ($0.5mv = 3.2 \times 10^{-32}$ *J*, if the mean particle size is 62 nm). Obviously, the kinetic energy of a particle in Brownian motion is significantly larger than that of the electrophoresis. Therefore, the main attractive force is contributed by the van der Waals force in this simulation case. Large particles exhibit a larger van der Waals force than fine particles. The sequential images show that particles in the cluster will flocculate and sometimes separate and rotate.

(a)

(b)

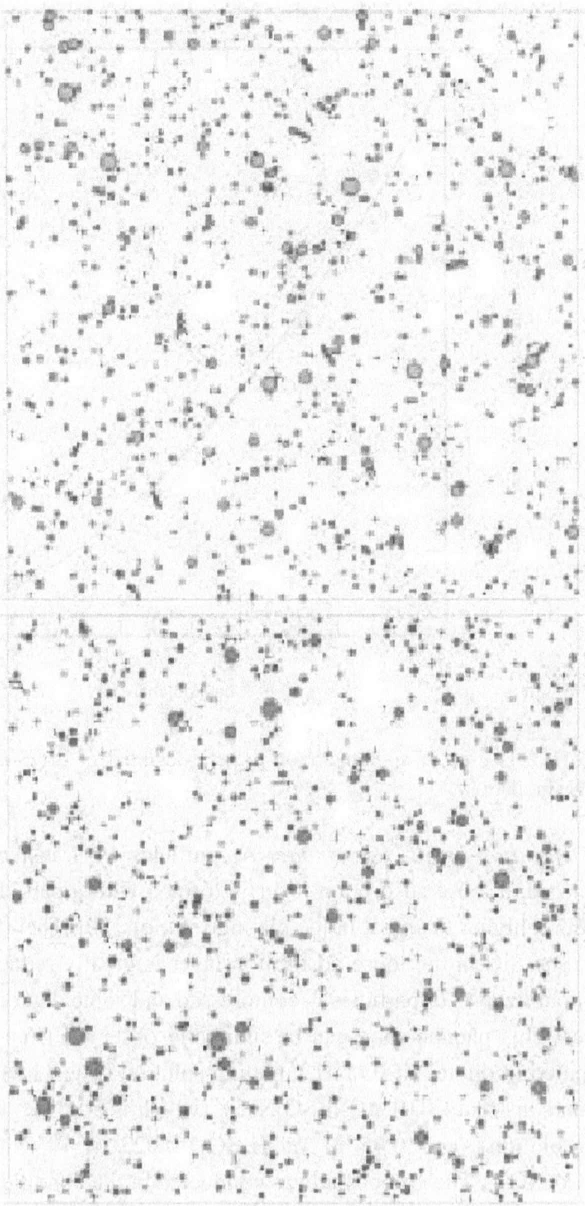

Fig. 11.2. DEM simulation of particles with ζ-potential either (a) 0 mV or (b) 30 mV

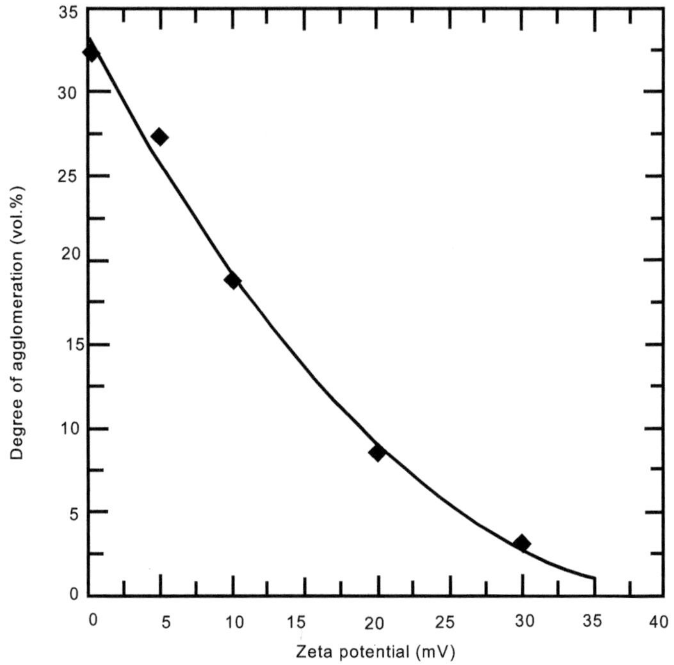

Fig. 11.3. Degree of agglomeration to zeta potential of the SiC particles determined by DEM simulation

Deposition of SiC to cathode: As particles are transferred to the OHP of the electrode, a loose adsorption layer is formed with a particle concentration ~20 to 60 times higher than that in the plating solution [20]. Therefore, the distance of the ultrafine SiC in the loose adsorption layer is greatly reduced. As a consequence, the double layer of particles is compressed and some degree of agglomeration may occur. This phenomenon can be indirectly observed from the composite coating prepared in dilute $Ni(SO_3NH_2)_2$ plating solution (Fig. 11.4). The dilute plating solution consists of 0.01 *M* $Ni(SO_3NH_2)_2$ (which is 50 times less than a normal plating solution), and 0.65 *M* of H_3BO_3, 0.023 *M* of $NiCl_2$, and 1×10^{-4} *M* of $Na_3Co(NO_2)_6$. The composite was conducted in the conditions of 0.1 vol.% of SiC, pH=5, at 4 ASD, 50°C for 1 h. An electric charging artifact occurs in the SEM micrograph due to a poor conductive surface. It implies that the content of electric conductive Ni is in very low concentration. The EDS analysis (Fig. 11.4b) shows that the SiC content is nearly 92 vol.% and Ni is less than 8%. The porous morphology is due to the packing situation of the loose SiC particle package.

When electrochemical reactions occur, the Ni^{+2} ions which have been adsorbed on the surface of SiC are reduced on the electrode. The particles are trapped by a thin Ni layer and fixed into the coating layer gradually. The SiC particles in loose adsorption change to strong adsorption, and form a SiC/Ni composite.

Fig. 11.4. (a) SEM micrograph of the surface of a SiC/Ni composite coating processed in a dilute (0.01M) $Ni(SO_3NH_2)_2$ plating solution, and (b) EDS analysis of (a)

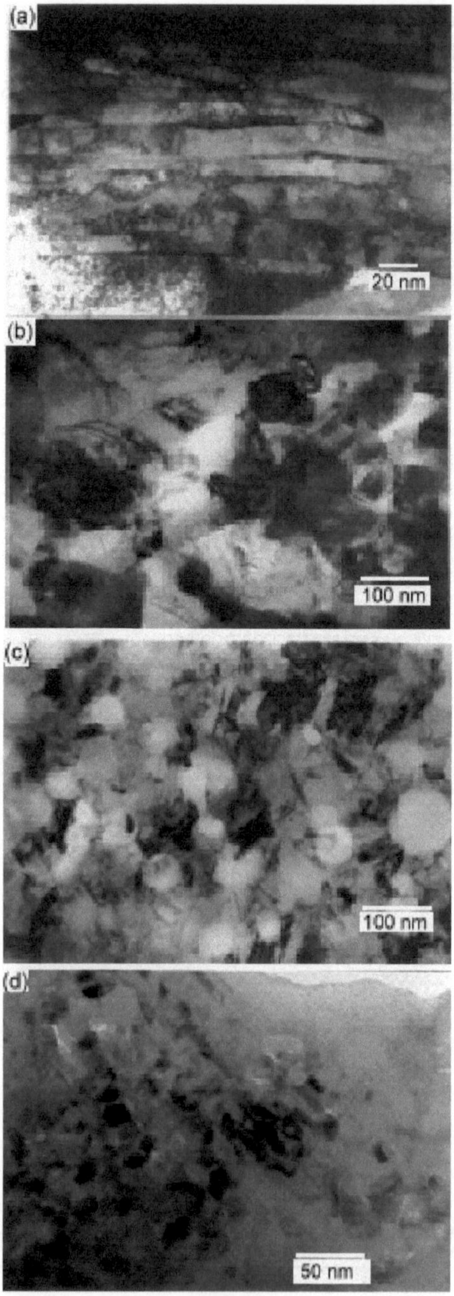

Fig. 11.5. TEM micrographs of the cross-section of (a) pure Ni, (b) SiC/Ni, (c) Al_2O_3/Ni and (d) ZrO_2/Ni composite layers

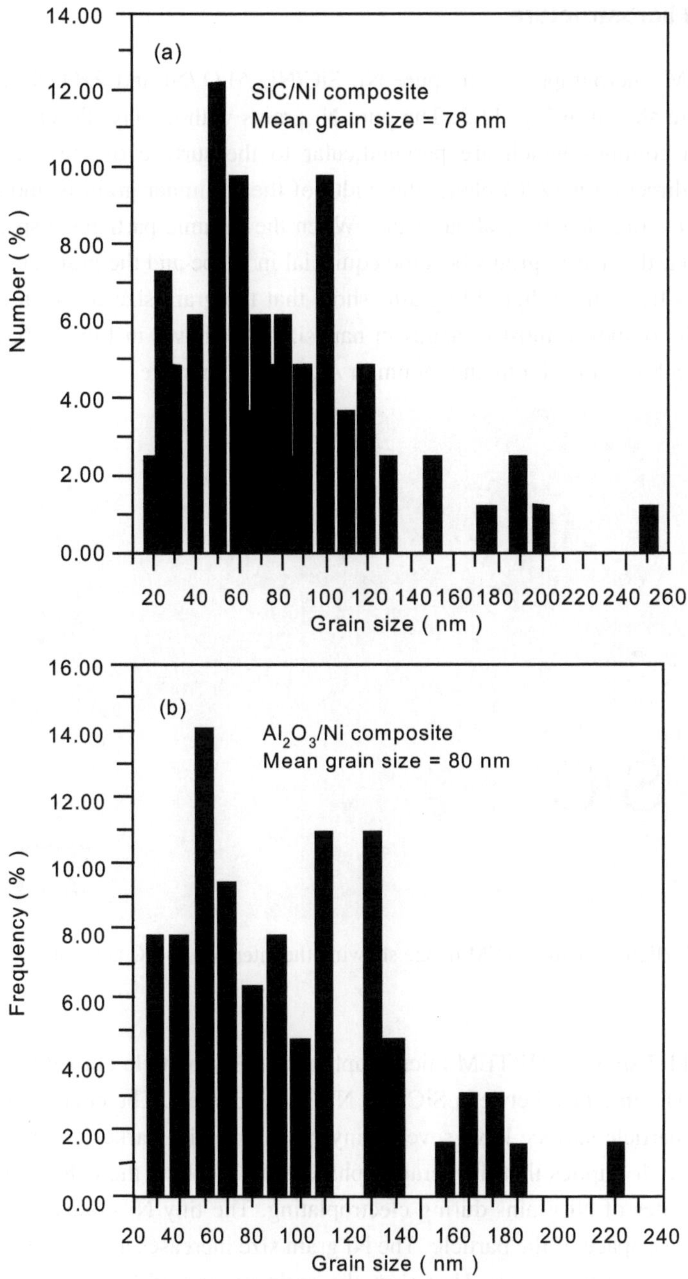

Fig. 11.6. The grain distribution of Ni in (a) ultrafine SiC/Ni and (b) ultrafine Al_2O_3/Ni composites

11.3.2 Microstructure

The TEM micrographs of the pure Ni, SiC/Ni, Al_2O_3/Ni and ZrO_2/Ni composite layers are shown in Fig. 11.5. The pure Ni grains without any dispersed particles grow in columns which are perpendicular to the surface of the cathode. The growth direction is (220) plane, the width of the columnar grain is about 0.2 μm and the average length is about 3 μm. When the ceramic particles disperse in the Ni matrix, the matrix grains become equiaxial in shape and the grain size reduces. Statistics from more than 200 grains show that the grain size of Ni in SiC/Ni or Al_2O_3/Ni composite mostly ranges in nanosize. As shown in Fig. 11.6, the mean Ni size in SiC/Ni is 78 nm and 89 nm in Al_2O_3/Ni composite.

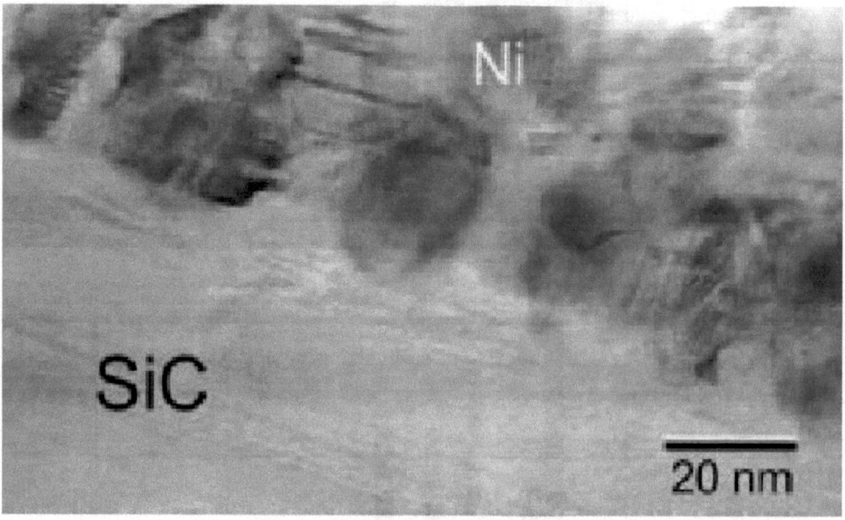

Fig. 11.7. High-resolution TEM image showing the interface of SiC and polycrystalline Ni grains

Fig. 11.7 shows a HRTEM micrograph of one SiC particle embedded in the Ni matrix. The interface between SiC and Ni is well contact. The concave regions of the SiC particle surface have several tiny Ni grains with darker contrast than the SiC phase. It implies that the ceramic phase can contribute the heterogeneous nucleation sites of Ni grains during electroplating. The tiny Ni grains grow and fill the concave space of the particle. The Ni grain size increases as the distance grows from the ceramic particles. However, the grain growth of Ni is along a specific preferred plane and forms a columnar texture, so that the pure Ni layer has a columnar microstructure. When ceramic inclusions are in high concentration, the growth of Ni is inhibited by inclusions, and creates equiaxial grains. The growth of Ni grains near two SiC particles is schematically illustrated in Fig. 11.8. The diagram shows that the Ni grains next to the SiC particles are equiaxial and small. The Ni grains become columnar and elongated as the distance increases from the SiC. As a result, a microstructure of columnar Ni grains radially grown from a SiC particle is possible. The microstructure may disappear as the amount of hard inclusion increases. Consequently, well-dispersed SiC particles in the SiC/Ni composite can result in equiaxial Ni grains in sizes of nanoscale.

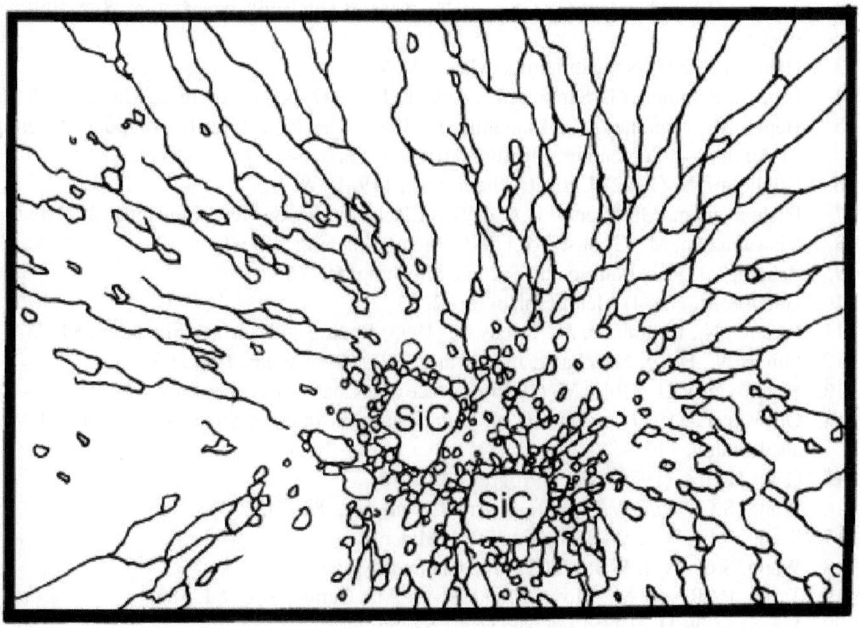

Fig. 11.8. Schematic diagram of grain growth of Ni in the composite plating

11.4 Conclusions

This study showed the microstructure and mechanical properties of Ni layers which were codeposited with ultrafine SiC, Al_2O_3, or ZrO_2 powder and formed ceramic/Ni composite layers. The grain shape of the Ni matrix is columnar for the pure Ni layer, but when the ceramic particles are dispersed in the matrix, the Ni grain changes from elongated grain (0.2 m× 3 m) to an equiaxial shape in an average size of 78 nm for SiC/Ni, and 50 nm for Al_2O_3/Ni layer. The concave region of the particle interfaces serves as the heterogeneous nucleation site of Ni grains.

The codeposition process of the ceramic particles on the plating solution can be classified into five stages: hydration of particles, adsorption of Ni^{2+} on particles, agglomeration of ultrafine particles, diffusion of the particles toward the cathode, and reduction and codeposition of particles with Ni to form composite layer.

References

1. Moore, D.F., Milne, W. I., Oda, S. (1999) Powder Engineering Journal, **4**, 89
2. Tompa, G.S., Skandan, G., Glumac, N., Kear, B.H. (1999) Am. Ceram. Soc. Bull., **78** [Oct.] 70
3. Hahn, H. (1997) Nanostructured Materials, **9**, 3
4. Erb, U., Palumbo, G., Szpunar, B., Aust, K.T. (1997) NanoStructured Materials, **9**, 261
5. Hepel, M., Mahoney, C., Tannahill, T. (1997) The 1997 Joint International Meeting, 31 August—5 September, Meeting Abstracts, Paris, 641
6. Guglielmi, N. (1972) J. Electrochem. Soc., **119**, 1009
7. Pushpavanam, M., Shenoi, B.A. (1974) Electroplating & Metal Finishing, **27**, 10
8. Viswanathan, M., Ghouse, M. (1979) Metal Finishing, **10**, 67
9. Zahavi, J. (1982) Plating and Surface Finishing, **69**, 76
10. Ghouse, M. (1984) Metal Finishing, **82**, 33
11. Periene, N., Èešuniene, A., Taicas, L. (1993) Plating and Surface Finishing, **80**, 73
12. Sun, K.N., Hu, X.N., Zhang, J.H., Wang, J.R. (1996) Wear, **196**, 295
13. Stankovic, V.D., Gojo, M. (1996) Surface and Coatings Technology, **81**, 225
14. Orowan, E. (1947) discussion in "Symposium on Internal Stresses", Institute of Metals, London, 145
15. Gahr, K.H.Z. (1979) Metal Progress, **116**, 46
16. Wang, S.C., Wei, W.J. (2001) J. Am. Ceram. Soc, **84** [7], 2001
17. Wang, S.C., Wei, W.J. submitted to J. Mater. Res.
18. Wang, S.C., Wei, W.J. (2003) Mat. Chem. Phy., (in press)
19. Tartaj, P., Reece, M., Moya, J.S. (1998) J. Am. Ceram. Soc., **81**, 389
20. Morrison, W.H. (1985) J. Coat. Tech., **57**, 55
21. Lewis, J.A. (2000) J. Am. Ceram. Soc., **83**, 2341
22. Fukuhara, M., Abe, Y. (1993) J. Mater. Sci. Lett., **12**, 681

12 Hot-Pressing Technique : SiC/AlN Composites

Y.-B. Pan, D.-L. Jiang, J.-H. Qiu, J.-K. Guo

12.1 Introduction

Silicon carbide (SiC) does not exist in nature; it must be synthesized by high temperature chemical reactions. The first commercial products made of silicon carbide appeared at the end of the 19th century when Acheson developed a process of reacting sand and coke in an Arc furnace. This is still the basic SiC manufacturing process in use today. Earlier, SiC was mainly used abrasive materials, because of its hardness originating from its covalent bonding. Since the 1970's, as the understanding of materials and fabrication techniques have improved significantly, SiC has been further considered for applications in structural ceramics. During this period, sintering methods and suitable sintering additives to SiC ceramics have been extensively explored.

Silicon carbide is a covalent bonded material; the crystal structures are either cubic or hexagonal lattice. Silicon and carbon are in tetrahedral coordination and the SiC structures are composed of layers of these tetrahedrals. All SiC structures are made up of a single basic unit, a plane of either SiC_4 or Si_4C tetrahedral [1]. The 3C polytype is cubic and designated as β-SiC, whereas hexagonal or rhombohedral forms (2H, 4H, 6H, 15R, 21R) are designated as α-SiC [1,2]. The physical properties of various SiC crystal are shown in Table 12.1.

- 2H is generally formed at a temperature below 1400°C. As the temperature rises there are successive changes in the polytypes from 2H to 3C, 4H and eventually 6H.
- 3C is often formed in the initial stage of crystal growth. At temperatures of 1600°C and above, it changes into α-type.
- At temperatures around 2000°C, 4H is the most possible phase.
- In the range of 2200°C—2600°C, 6H is more stable than 2H, 3C, 4H and 15R.

12.2 Applications of Silicon Carbide Ceramics

Dense silicon carbide-based ceramics can be fabricated by pressureless sintering, hot-pressing, reaction-bonding, chemical vapor deposition (CVD), and repeated impregnation of polycarbosilane into the porous SiC compact.

During the reaction-bonding, little or no shrinkage occurs in the fabrication of SiC. The technique can thus be applied to the fabrication of complex shapes or high-precision components and large components. The additive for pressureless sintering, the B-C system and the Al-C system are widely used.

Hot-pressed silicon carbide can be fabricated to essentially full density and high strength by using various additives. These sintering additives allow hot-pressing at a lower temperature. CVD can produce SiC with complex shapes, although the technique may be too expensive for production.

Silicon carbide is an attractive candidate material for many structural applications, because of its raw material availability in powder form, price for competitiveness, good specific strength, excellent environmental properties, and shape forming ability. In particular, it shows excellent resistance to oxidization, erosion, corrosion and thermal shock. A high-temperature strength retention of up to 1500°C has also been demonstrated. For these reasons, it has been widely used in applications as a wear- and erosion-resistant component in the chemical and automotive industries. Many examples are given in Table 12.2.

Therefore, silicon carbides are extensively used in wear application at low temperatures and have also been used for several high-temperature applications [5,6]. In early 1980's, SiC ceramics became excellent candidate material for heat engine components. However, its low strength, reliability and fracture toughness imposed limitations for many applications.

12.3 SiC/AlN Particulate Composites

The fracture toughness and strength of SiC at room temperature are rather low. These disadvantages prevent the practical use of silicon carbide as a high performance structural material. On the other hand, because SiC is a covalent bonding compound, there are some problems associated with this bonding characteristic. (1) SiC must be sintered at a temperature above 2000°C. (2) The sintering of SiC is difficult without using suitable additives. (3) SiC grains sintered at high temperature are easily coarsened and result in low mechanical properties. Thus, the study of SiC ceramics focuses on three problems as follows: (1) selecting a

Table 12.1. The physical properties of SiC materials [3]

Name	Silicon carbide (Moissanite)				
Molecular formula	SiC				
Formula weight	40.1				
Crystalline	3C (Cubic)	2H (Hexagonal)	4H (Hexagonal)	6H (Hexagonal)	15R (Rhombohedral)
Shape	Sphere		Multi-hexagonal	Cylinder	Sphere
Lattice	a_0=0.4359	a_0=0.3079	a_0=0.3079	a_0=0.3082	a_0=0.3079
Parameter [nm]		c_0=0.50480	c_0=1.0254	c_0=1.5116	c_0=3.778
Optical parameter	pl. axes	Axial (+)			
		ξ=2.697			ω=2.654
Density [(g/cm^3]	3.10		3.028		
Decomposition temperature [°C]	~2830				
Elastic modulus [GPa]	390-460				
Thermal expansion Coef. [1/°C]	4.4-4.9×10-6 (25-2000 °C)				
Shore hardness	9.5				
Knoop hardness [MPa]	281.5, 3C⊥(111)				
Specific heat [J/g. °C]		0.712	(0°C)		
		0.921	(200 °C)		
		1.172	(1000 °C)		
		1.256	(1400-2000 °C)		
Thermal conductivity [W/m.K]		502.3	(0 °C))		
		62.8	(700 °C))		
Bandgap [J]	3.52×10-19 (3C)				
	4.58×10-19 (6H)				
	5.28×10-19 (2H)				
Electron mobility [cm^2/V.sec]	1000 (3C)				
	500 (15R)				
	200-300(6H)				
Debye temperature [°C]	1157				
Compressibility [MPa]	1.2-2.1×10-8				

Table 12.2. Examples the application of silicon carbide materials [3-10]

Field	Application condition	Applications	Major advantages
Petroleum industry	High-temperature and high-pressure (liquid) abrasive	Nozzles, Bearing, Sealing, Valve	Wear resistance, heat consumption
Micro-electric industry	High power radiating	Sealing materials, Substrate	High heat conductivity
Chemical industry	Strong acid or alkali, and high temperature	Sealing, Bearing, Pump sleeve, Heat exchanger, gasification tube, thermocouple	Wear resistance, corrosion resistance, Gas-impermeable resistance, Oxidization resistance
Automobile, tractor, aircraft, space rocket	Combustion	Combustion parts, turbine booster, turbine blade, Gas-combustion engine blade, Rocket nozzle	Low friction, high strength at high temperature, Low inertia load, heat shock resistance
Laser	High power and temperature	Reflecting screen	High stiffness and stability
Sand blower	High speed abrasion	Nozzle	Wear resistance
Paper industry	Pulp refuse (50%NaOH)	Sealing sleeve, bearing liner	Corrosion resistance, Wear resistance, low friction
Steel industry	High temperature gas	Thermocouple insulating tube, Radiating tube, Exchanger	High temperature resistance, Corrosion resistance, Gas-impermeable
Mining	Abrasion	Liner, Pump parts	Wear resistance
Atomic energy	B-contained high temperature water	Sealing, and sleeve	Radiant resistance
Others	Plastic industry	Wire-drawing, Die forming	Wear resistance, corrosion resistance

suitable second phase, (2) improving the toughness, and (3) decreasing the sintering temperature.

Recently, silicon carbide matrix composites with improved fracture toughness and increased strength have been developed [11—17] by incorporating some non-oxide particulates such as TiC [18—20], TiB_2 [21], ZrB_2 [22] or oxide as Al_2O_3+Y_2O_3 [23]. SiC-AlN composites were also developed some years ago. The important step in the design of a SiC matrix composite is the selection of the second phase particle and the formation feasibility of SiC-AlN composites. Since Cutler and Miller found that 2H polymorph of SiC and AlN can form solid solution [24], much of the research has focused on a SiC-AlN binary system [25,26]. The strength and toughness of the SiC-AlN solid solution [27—29] were quite low, but it is interesting that AlN was found to inhibit the grain growth of SiC [27]. SiC-AlN particulate composites were previously fabricated by mechanical mixing of the SiC and AlN powders, and then hot-pressed at 1950°C under 40 MPa for 0.5 hr [26]. A fully dense composite with partial solid solution and steep compositional gradient is obtained.

12.3.1 Microstructure

During the sintering process, SiC and AlN will be vaporized at high temperature, and produce a deposit on the surface of particles. The rate of transport from the surface is proportional to the equilibrium vapour pressure over the surface. The solid vaporization pressures of AlN and SiC reach 14 and 1.2×10^{-3}mmHg at 1900°C, respectively [30].

Based on the observation of the interface between SiC grains [31] by TEM and EDS, the system of SiC-AlN exhibits three stages to the formation of the solid solution. First, AlN particle is vaporized. AlN is then deposited onto the surface of SiC grains, accompanied by the reduction of AlN particle size. Finally, partial SiC and AlN solid solution is formed at the SiC grain boundaries. The microstructure of AlN/SiC composite is shown in Fig. 12.1.

12.3.2 Microhardness

For microhardness testing, a hard indentor of specified geometry is pressed the into surface. The pressed surface is deformed by vertical force with sliding or scratching, and as a result the damage takes place in a localized region, which is very sensitive to the surface condition. Therefore, microindentation hardness is widely used for materials evaluation. The microhardness of the SiC-AlN composite specimen decreases from 2351 to 1049 with increasing temperature, as shown in Fig. 12.2. The SEM micrographs of the microhadness indentation made at 25°C and 1200°C are shown in Fig. 12.3.

Indentation hardness decreases as the holding time is increased, and decreases with the increase of temperature. This is probably due to creep deformation and the decrease of elastic modulus at elevated temperature.

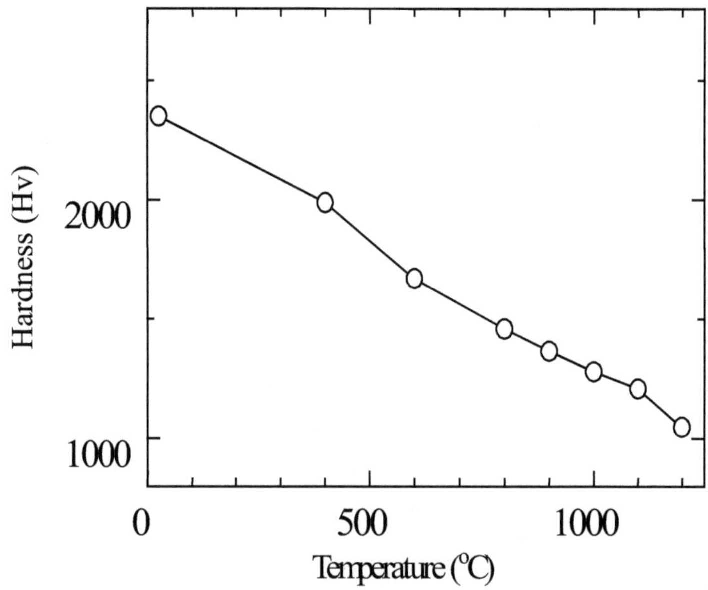

Fig. 12.1. The microhardness of SiC-AlN system as a function of temperature

12.3.3 Mechanical Properties

The fracture toughness of AlN/SiC composites hardly changes from room temperature to 1400°C, as shown in Fig. 12.4. Due to the formation of SiC-AlN solid solution at the boundary of the SiC grains, compressive stress exists within the SiC grains. The SiC-AlN solid solution is stable at high temperature, the mechanical properties are maintained till 1000°C. The mechanical properties are lower above 1000°C. This may be due to (1) a weakening of the boundaries and (2) the softening of SiC-AlN as shown in Fig. 12.2.

(a)

(b)

Fig. 12.2. The SEM micrographs of the AlN/SiC composite indented at (a) 25°C and (b) 1200°C

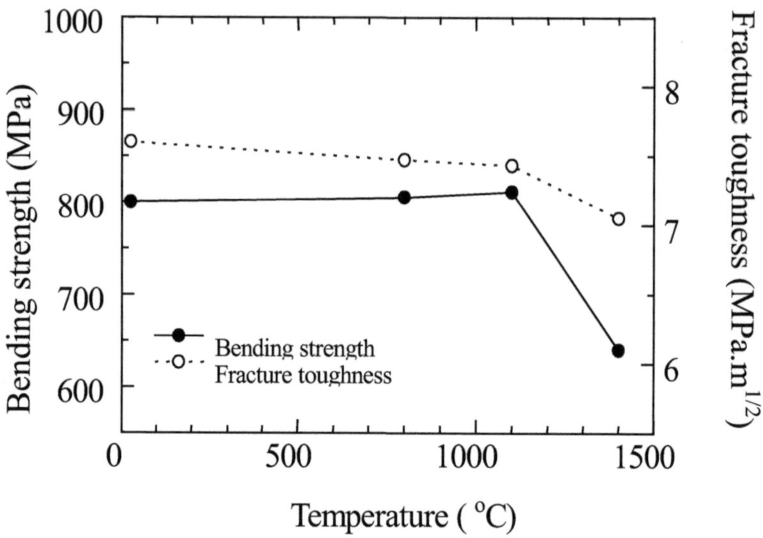

Fig. 12.3. The mechanical properties of AlN/SiC composite

12.4 Conclusions

Silicon nitride is one of the most stable materials at elevated temperatures. SiC can combine with AlN to form a high-temperature material. The SiC/AlN composite can be prepared successfully by hot-pressing. The potential of the applications of SiC/AlN composites at elevated temperature is high.

Acknowledgements

The authors thank the support from the 973 project (G2000067200) and NSFC.

References

1. Shaffer, P.T.B. (1969) Acta Cryst, **B25**, 477
2. Bind, J. (1978) Mat. Res. Bull., **13**, 91
3. Inomata, Y. (1988) SiC Ceramics, ed. by Shigeyuki, M., Inomata, Y., Uchida, M., Rokakuho Publishing Co., Ltd., Japan, 1

4. Saito, S. (1988) Fine Ceramics, Elsevier Ohmsha, Ltd., Tokyo
5. Leatherman, G.L., Katz, R.N. (1989) in Superalloys, Supercomposites, and Superceramics, Academic Press Inc. 671
6. Larson, D.C. (1979) Technical Report AFML-TR-79-4188
7. Dutta, S. (1985) J. Am. Ceram. Soc., **68,** c-269
8. Weimer, A.W. (1977) Carbide, Nitride and Boride: Materials Synthesis and Processing, Chapman & Hall
9. Ault, N.N., Crowe, J.T. (1991) Ceramic Bulletin, **70**, 881
10. Freedman, M.R., Millard, M.L. (1986) Ceram. Eng. Sci. Proc. **6**, 884
11. Jiang, D.L. (1995) J. Inorganic Materials, **10**, 151 (in Chinese)
12. Evans, A.G. (1990) J. Am. Ceram. Soc., **73**, 187
13. Garnier, J.E. (1987) Am. Ceran. Soc. Bull., **66**, 325
14. Niihara, K., J. Japan. Ceram. Soc., **99**, 974
15. Kodama, H., Miyashi, T. (1990) J. Am. Ceram. Soc., **73**, 3081
16. Kim, D-H, Kim, C-H (1986) J. Am. Ceram. Soc., **11**, 824
17. Morita, M., Kinbara, I., Fukuda, H. (1988) Composites, Nikan Industry News Publisher, Japan
18. Wei, G.C., Becher, P.F. (1984) J. Am. Ceram. Soc., **67**, 571
19. Endo, H., Ueki, M., Kubo, H. (1991) J. Mater. Sci., **26**, 3769
20. Jiang, D.L., Wang, J.H., Li, Y.L., Ma, L.T. (1989) Mater. Sci. Eng., **A109**, 405
21. McMurtry, C.H., Boecker, W.D., Seshadri, S.G., Zanghi, J.S., Garnier, J.E. (1987) Am. Ceram. Soc. Bull., **66**, 325
22. She, J.H., Jiang, D.L., Greil, P. (1991) J. Europ. Ceram. Soc., **7**, 243
23. Larker, H.T., Hermansson, L., Adlerborn, J. (1988) Industrial Ceram., **8**, 17
24. Cutler, B., Miller, P.D., Rafaniello, W. (1978) Nature, **275**, 434
25. Rafaniello, W., Plichta, M.R., Virkar, A.V. (1983) J. Am. Ceram. Soc., **64**, 272
26. Zangil, A., Ruh, R. (1988) J. Am. Ceram. Soc., **71**, 884
27. Raffaniello, W., Cho, K., Virkar, A.V. (1981) J. Mater. Sci., **16**, 3479
28. Ruh, R., Avigdor, R. (1982) J. Am. Ceram. Soc., **65**, 260
29. Lee, R-R, Wei, W-C (1990) Ceram. Eng. Sci. Proc., **11**, 1094
30. Camcohob, T.B. (1965) Handbook of High-melting Point Compound, China Industry Publisher, China
31. Pan, Y., Qiu, J., Morita, M. (1998) J. Mater. Sci., **33**, 1233

13 Nondestructive Evaluation of Strength and Residual Stress using Sphere Indentation

Y. Bao, L. Sun

13.1 Introduction

Hertzian indentation is a convenient test method and has been extensively used in the characterization of fracture and deformation properties of brittle materials [1,2]. Indentation damage study and its applications have lately drawn more and more attention [3,4], and extend to many practical areas, such as bearings, ceramic coating, dental restoration [5] and residual stress evaluation [6]. For brittle materials, the initiation of a Hertzian cone crack represents the beginning of indentation damage. The critical indentation stress for cracking on the surface is an important parameter for the analysis of strength for brittle materials. However, many contradictions exist in evaluating strength through sphere indentation [7,8]. The key issues to be solved are as follows:

(i) The maximum indentation stress for crack initiation is usually much higher than the bulk strength, and decreases with increasing sphere radius.

(ii) The indentation crack is always larger than the contact circle, i.e. the crack does not initiate from the boundary of the contact area where the maximum stress is applied.

These phenomena are contradictory to conventional strength theory and indicate that crack initiation in a brittle solid does not depend on the value of peak stress. In fact, the critical stress state varies from material to material; a general mode suitable for all materials is almost impossible. However, when material and sphere indenter are given, the critical stress state for Hertzian cracking obeys a unique law, which will be discussed in the sections below. This study tries to give an explanation for the above-mentioned paradoxes, through the effect of stress gradient on strength. The focus is placed on exploring the solutions to the following questions: i) Why does the Hertzian crack in brittle materials not initiate from the position of maximum stress? ii) What is the critical state for contact cracking in brittle materials to take place? iii) How can we evaluate the inert strength of a glass material nondestructively? Furthermore, we aimed to develop a methodology to determine the local strength and residual stress in glass by using sphere indentation.

13.2 Effect of Stress Gradient on Fracture

Hertzian contact theory for a sphere with radius r, at normal load P, on a flat continuum elastic body is well known [9,10]. The contact radius is given by

$$a^3 = 4kP \cdot r/(3E) \tag{13.1}$$

where $k=(9/16)[(1-\nu^2)+(1-\nu'^2)E/E']$, E, E' and ν,ν' are Young's modulus and Poisson's ratio for the specimen and the sphere indenter. The mean pressure in the circle contact area is p_o.

$$p_o=P/(\pi a^2) \tag{13.2}$$

Combining with (13.1), (13.2) can be written [11] in terms of P and r

$$p_o = \left(\frac{3E}{4k}\right)^{2/3} \cdot \frac{1}{\pi} \cdot \left(\frac{P}{r^2}\right)^{1/3} \tag{13.3}$$

Crack initiation in brittle materials is mainly caused by tensile stress. The ring crack then results from the radial stress σ_R in the Hertzian elastic contact stress field. In axis-symmetric cylindrical coordinates R, z, the radial stress σ_R is given by [11]

$$\frac{\sigma_R}{p_o} = \frac{1}{2}(1-2\nu)(\frac{a}{R})^2\left[1-\left(\frac{z}{\sqrt{u}}\right)^3\right] + \frac{3z}{2\sqrt{u}}\left[\frac{(1-\nu)u}{(a^2+u)} + (1+\nu)\frac{\sqrt{u}}{a}\arctan\left(\frac{a}{\sqrt{u}}\right) - 2\right] \tag{13.4}$$

where $u = \frac{1}{2}\left\{(R^2+z^2-a^2) + \left[(R^2+z^2-a^2)^2 + 4a^2z^2\right]^{1/2}\right\}$

The maximum tensile stress occurs at the boundary of the contact area (z=0, R=a).

$$\sigma_m = \frac{1}{2}(1-2\nu)p_o \tag{13.5}$$

Stress calculation shows that the radial stress at contact circle rapidly degrades and becomes compressive along the depth, i.e., stress gradient at the contact circle along depth direction is very steep, as shown in Fig. 13.1. The contact stresses exhibited these special features: (a) the stress gradient along depth decreases with increasing contact radius (or sphere radius); (b) the stress gradient along depth becomes smoother with increasing distance to the contact circle R/a, and the steepest stress gradient is at R/a=1, as shown in Fig. 13.1.

Many experiments indicate that, for brittle materials in highly uneven stress field, fracture dose not certainly occur at the point of peak stress. The maximum stress to fracture could be much higher than the bulk strength. Under a stress intensity factor below K_{IC}, for example, a brittle material does not fracture although the stress at the crack tip may be much higher than the tensile strength of the material. In those cases, the strength (peak stress at fracture) will be variable with stress distribution, not a material constant. In fact, the critical state for crack initiation in brittle materials depends on the mean stress in a specific small zone (process zone), rather than on the peak stress at one point. This principle was termed the mean-strength criterion [12]. When the average stress in the process zone attains a critical value σ_o, the brittle solid will crack in this region. The width of the process zone, Δ, is a material constant dependant on the properties of material but independent of specimen size and configuration. σ_o is the inert strength of materials. Thus, the condition for the initiation of ring crack under indentation load is given by:

$$\frac{1}{\Delta}\int_0^{\Delta} \sigma_R \, dz = \sigma_o \tag{13.6}$$

The value of Δ is related to the material properties as,

$$\Delta = \frac{2}{\pi}\left(\frac{K_{IC}}{\sigma_o}\right)^2 \tag{13.7}$$

Generally, the process zone size Δ is an increasing function of grain size for brittle materials. The process zone size of glass and fine-grained ceramics is around 0.03—0.04 mm. The mean-strength criterion indicates that the critical peak stress in uneven stress field is not constant, but depends on the stress gradient. The steeper the stress gradient is, the higher the critical peak stress will be. It indicates that the critical maximum stress in the contact stress field would increase with decreasing sphere radius because the stress gradient increases with decreasing sphere radius. Although the maximum tensile stress in the contact stress field occurs at the boundary of the contact zone, the maximum mean stress is at the outside of the contact zone, as shown in Fig. 13.2. The size of the ring crack, therefore, should be greater than the size of the contact zone in the light of the mean-strength criterion.

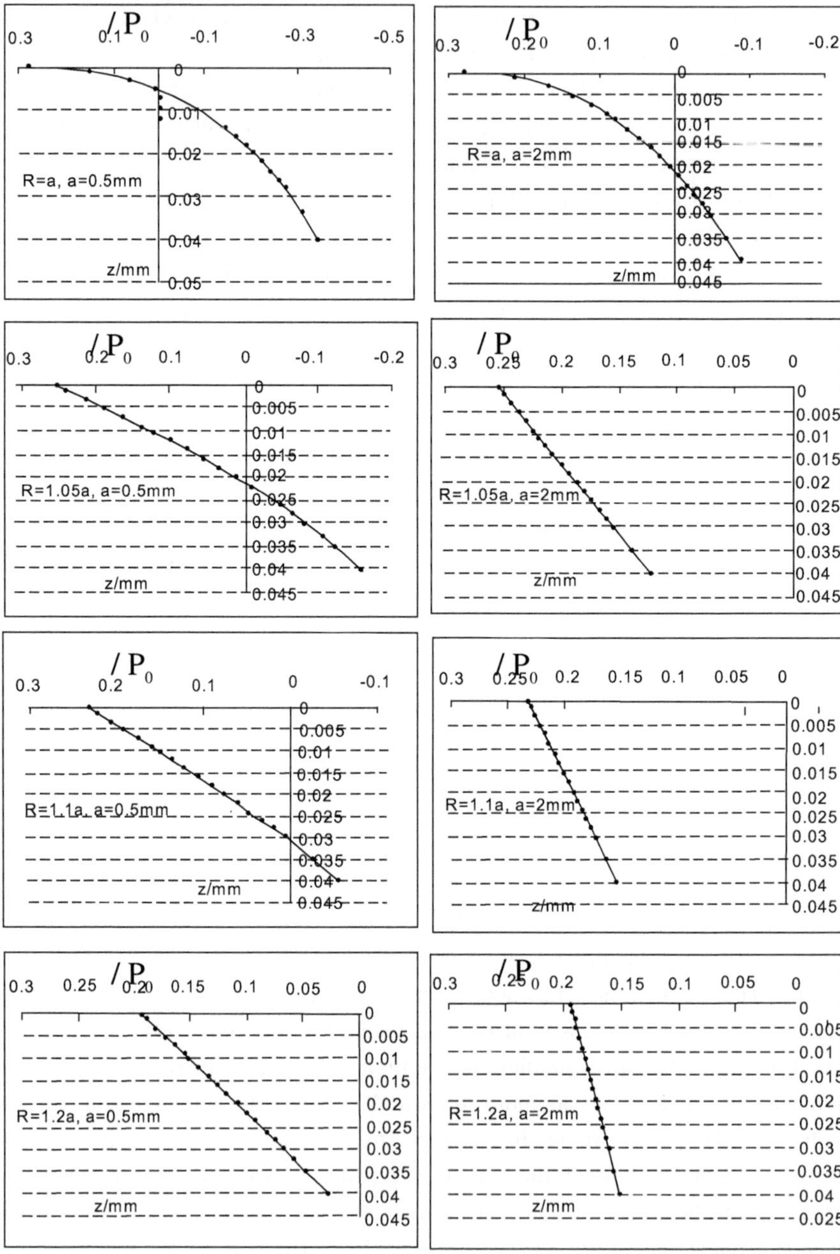

(a) series for a=0.5mm (b) series for a=2mm

Fig. 13.1. Calculated radial stresses outside the indentation area

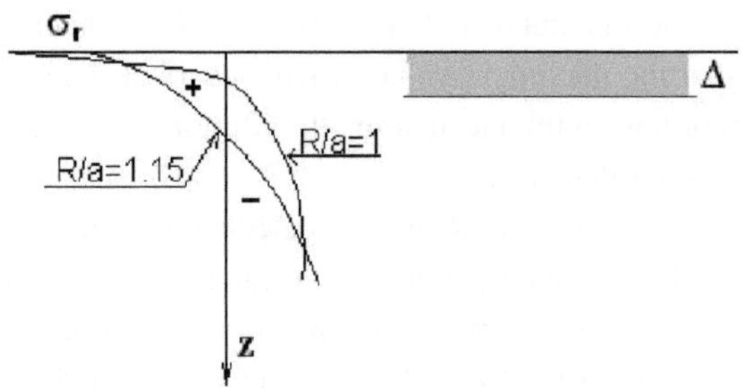

Fig. 13.2. Schematic of radial stress distribution along depth at different radial coordinates (R=a and R=1.15a)

13.3 Critical Issues and Local Strength

As an example, an alumina spherical indenter and a soda-lime glass specimen are considered in the calculation and tests. Young's modulus and Poisson's ratio are respectively 310 GPa and 0.20 for the alumina sphere, 72 GPa and 0.22 for the glass specimens. The mean stress in the process zone, s, is a function of R

$$s(R) = \frac{1}{\Delta}\int_0^{\Delta} \sigma_R(z,R)dz \tag{13.8}$$

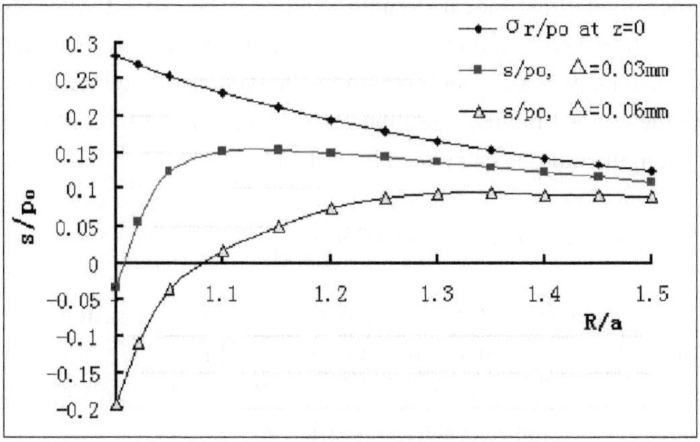

Fig. 13.3. Surface radial stress and mean stress as a function of R/a for Δ=0.03 and 0.06 mm

Based on radial stress distribution and numerical integration of (13.8) and (13.4), we obtained the mean stress as a function of R/a, as shown in Fig. 13.1. It can be seen that the maximum radial stress is at R=a where the mean stress is minimum and that the maximum value of s/p_o occurs at R/a=1.15 for the case of Δ=0.03mm. According to the mean-strength criterion, the ring crack should initiate at R/a=1.15, rather than at R=a.

The maximum mean stress at R_o/a is s_m, calculated by numerical integration for various contact radius (or sphere radius). The calculated results indicate that, with the increase of sphere size, the ratio s_m/p_o enhanced nonlinearly and approximately has a general form as in (13.9) in the range of most practical indentation sizes (0.2<a<0.6mm).

$$s_m = \left[C_1 + C_2 a + C_3 a^2\right]\cdot p_o \tag{13.9}$$

where C_1, C_2, C_3 are the constants related to the material properties. The maximum mean stress to crack initiation, s_m, represents the local strength of the specimen, σ_{loc}, at the place of indentation. It can be also considered as the inert strength. For the glass with ~ 0.03 mm of process zone size, combining (13.2), the local strength has approximately the form:

$$\sigma_{loc} = \left[-0.1373 a_c^{\,2} + 0.2862 a_c + 0.0236\right]\cdot \frac{P_c}{\pi a_c^2} \tag{13.10}$$

where a_c is the radius of contact zone in critical state. Since the radius of contact area a_c is difficult to accurately measure but the radius of ring crack R_o is easy to obtain after unloading, the former is sometimes approximately replaced by the latter for the calculation of local strength. Note that the contact radius estimated by this way may be 10~20% larger than the real value, because R_o is larger than a_c. In order to calculate the local strength more conveniently and to avoid errors in measuring the contact radius, when the elastic constants of the sample are known, the local strength could be calculated by (13.11).

$$\sigma_{loc} = -\left[0.1373 a_c^2 + 0.2862 a_c + 0.0236\right]\cdot\left(\frac{3E}{4k}\right)^{2/3}\cdot\frac{1}{\pi}\cdot\left(\frac{P_c}{r^2}\right)^{1/3}, \tag{13.11a}$$

where a_c can be determined by (13.1) with $P=P_c$. For a tungsten carbide sphere (r=2.5 mm) with Young's modulus of 600 GPa and Poisson's ratio of 0.25, and a glass specimen with Young's modulus of 72 GPa and Poisson's ratio of 0.22, the local strength can be approximately simplified as

$$\sigma_{loc} = 8.21 P_c^{1/3} + 3.014 P_c^{2/3} - 0.0437 P_c \tag{13.11b}$$

Since the strength of strengthened glass equals the sum of original strength and the residual compressive stress, the residual stress is estimated approximately by the difference between the local strength of the glass specimens after and before being strengthened, i.e.,

$$\sigma_r = \sigma'_{loc} - \sigma_{loc}, \tag{13.12}$$

where σ_{loc} is the local strength of the strengthened glass.

The relationship between the local strength and bulk strength can be estimated by using the size effects of strength or by experimentation. As for a known material and a given spherical indenter, the ratio, σ_{loc} / σ_b , should be a constant. Note that this constant is different from material to material. Thus, the inert strength of brittle materials can be simply determined by the indentation method. Assuming considering the local strength represents approximately the strength of one unit volume (: 1 mm^3), the bulk strength, σ_f, can then be estimated by using Weibull statistical theory

$$\sigma_f = \left[\frac{V_o}{V}\right]^{1/m} \cdot \sigma_o = \frac{\sigma_o}{V^{1/m}}, \tag{13.13}$$

where V is the volume of the specimen, m is Weibull modulus of strength. Generally, a bulk specimen has a volume larger than the unit volume, but a fiber specimen may not. So the strength of a glass fiber is usually near or even higher than the level of the inert strength.

13.4 Experimental Work

The local strength of soda-lime glass before and after strengthening was measured, using 10 mm thick plate the specimens. Loading rate was 0.5 mm/min. A ceramic sphere with a radius of 6 mm was used in the tests. It is important to determine the critical load Pc under which the ring crack is initiated. In this study, Pc was determined simply by observation with microscope during step loading, increasing or decreasing the applied load according to the observed results. The ring crack induced by critical load differs from the cone crack yielded by an overload, and the difference is shown in Fig. 13.4. The diameter of the ring crack, a_c, was measured by precision microscope. The experimental results for non-strengthened and strengthened glasses are shown in Table 13.1.

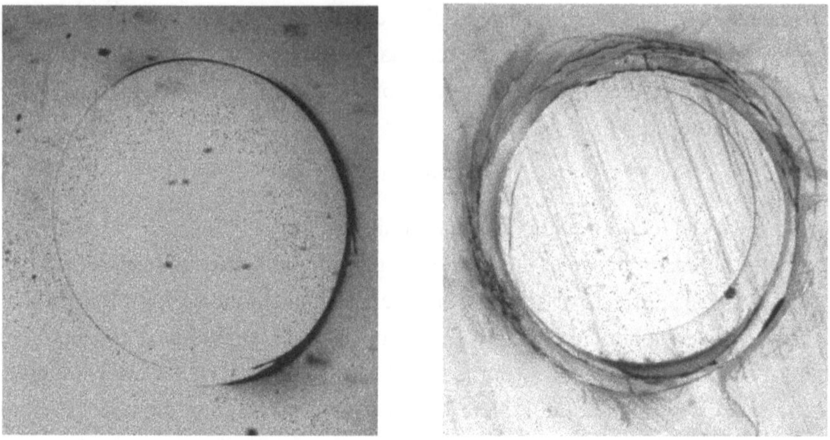

Fig. 13.4. (a) The ring crack on glass under the critical load. (b) The indentation crack under the load that is over the critical load

Note that the local strength is higher than the bending of bulk specimens. By comparing the local strengths between non-strengthened glass and strengthened glass specimens, it is easy to estimate the residual stress (compressive) on the surface layer of the strengthened glass. The residual stress σ_r is also the mean stress in the process zone; the actual residual stress has a steep gradient along depth from surface for the strengthened glass. In this study, the residual stress in the strengthened glass is 223•112=111 MPa, which is in good agreement with the results measured using polarizing technique.

The experiments indicated that the critical load to ring crack is proportional to the radius of spheres, and the mean pressure in the contact area decreases with increasing sphere size. Therefore the peak stress at the boundary of contact circle will decreases with increasing sphere size according to (13.5). Experiments show that the local strength obtained form the indentation tests is almost a constant for different spheres. Steel spheres with various radiuses were also used in the tests to measure the critical load P_c. The experimental results for glass and strengthened glass (different from the glass in previous tests) are shown in Fig. 13.5. The value of Pc/r is almost a constant that agrees with Auerbach's law.

13.5 Proof Tests of Strength

Proof tests are nondestructive tests to check whether the strength of a component is higher or lower than the expected level. Hertzian contact may be the simplest method proof test. Given an expected strength level, apply a contact load to

Table 13.1. The experimental data of sphere indentation with alumina sphere (r=6 mm) on glass and strengthened glass specimens

No.	Glass specimens (10 mm thick)		
	P_c [kg]	$2a_c$ [mm]	σ_{loc} [MPa]
1	118	1.13	111
2	112	0.96	132
3	117	1.15	107
4	107	1.04	113
5	120	1.16	109
6	115	1.12	110
7	114	1.12	108
8	115	1.15	106
Mean value			112 MPa
Standard deviation			8.3 MPa

No.	Strengthened glass specimens (10 mm thick)		
	P_c [kg]	$2a_c$ [mm]	σ_{loc} [MPa]
1	260	1.20	224
2	257	1.25	209
3	261	1.18	230
4	264	1.25	215
5	260	1.15	238
6	261	1.19	228
7	262	1.20	226
8	263	1.25	214
Mean value			223 MPa
Standard deviation			9.8 MPa

make the maximum mean stress in the process zone equal to the expected strength, and then check whether the ring crack has yielded. If the ring crack is induced, the strength of the material is lower than the expected strength; if no ring crack occurs, the strength is higher than the expected level. In this way, no special testing machine or standard specimens are required for proof test. It is a convenient testing method to examine the strength variation of brittle components in application. For example, by knowing the maximum working stress in a glass windshield is 50 MPa, if the safe factor is 3, then the strength of the windshield should be higher than 150 MPa. How to examine the safety of the windshield? Based on the calculation by (13.10) for steel sphere of 3 mm radius, the mean stress of 150 MPa corresponds to about 700 N indentation load. If 700 N load is applied and no ring crack is induced, then one can believe that the strength of the windshield is higher than expected.

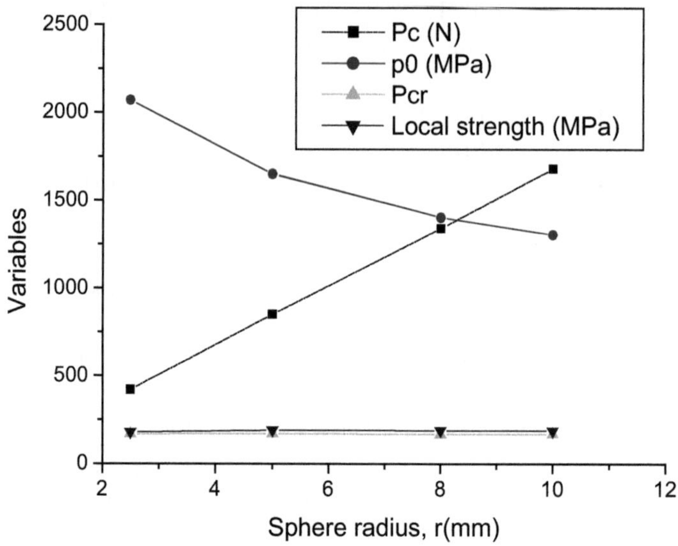

(a) Non-strengthened glass

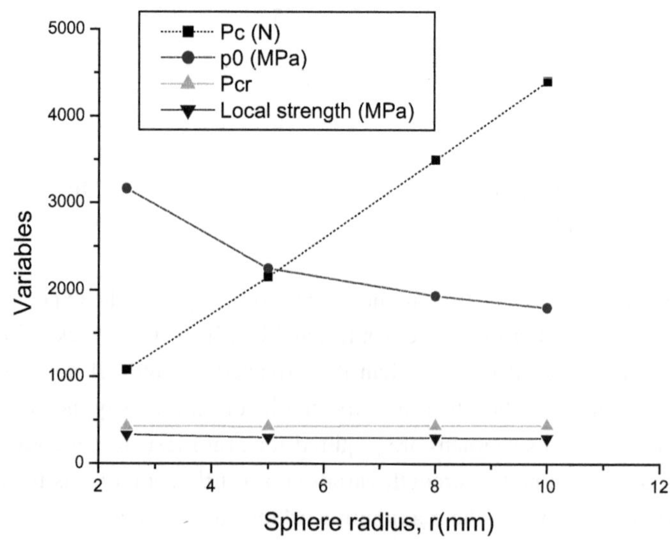

(b) Strengthened glass

Fig. 13.5. Experimental results of critical load, indentation stress, ratio of Pc/r and the local strength versus sphere radius for non-strengthened and strengthened glass (10 mm thick). Steel spheres were used

General proof test procedure is described as follows: (i) give a strength requirement S; (ii) calculate the indentation load P_S at which the maximum mean stress (local strength) equals S; (iii) apply an indentation load on the sample up to Ps, and check whether a Hertzian crack is induced on the contact point after unloading, to decide whether the local strength of the material is higher or lower than the expected strength S.

The proof test can be used for (a) choosing n samples with the strength higher than a known value from n+m samples; (b) checking the lower limit of allowing strength or residual strength; (c) distinguishing whether the surface strength of a component is higher than a given value.

13.6 Discussions and Conclusions

The unavailability of conventional strength criteria for indentation cracking in brittle materials is ascribed to the great stress gradient near the indentation surface along depth. Stress analysis and experimental results demonstrated that the initiation of Hertzian cracks in brittle materials depends on the mean stress in the process zone, so the peak stress in critical state is not constant for different stress gradients. The steeper the stress gradient is, the higher the critical peak stress will be. Thus, the critical peak stress does not reflect the local strength. The calculation shows that the mean stress attains the maximum at $R=R_0$ ($R_0>a$) and the minimum at the contact radius R=a where the maximum tensile stress occurs. That is probably the reason why the size of the Hertzian ring-crack is always greater than the size of the contact area for brittle materials tested by spherical indentation. The calculation based on the mean-stress criterion confirmed that the ratio of the crack radius to the contact radius, R_0/a, depends not only on the material properties (the process zone) but also on the sphere size. For sphere sizes in the range of r=2—8 mm and glass samples, the ratio R_0/a varies from about 1.3 to 1.1.

The local strength is reflected the inert strength of the materials and the upper limit of the bulk strength. It could be approximately considered as the strength of unit volume. Therefore the local strength is higher than the bulk strength. The bulk strength can be estimated with volume effect of strength for brittle materials. In fact, bulk strength of glass is influenced by many factors, such as sample sizes and shapes, testing methods, flaw distribution, and so on. In contrast, the local strength is independent of those factors. For example, the local strength will not be influenced by cracks at sample edges. Additionally, when the local strength is calculated with (13.10), the error from the measurement of contact size can be eliminated. That is why the standard deviation of local strength of glass is smaller than that of bulk strength. Obviously, the local strength can be considered as a material constant and it is proportional to the bulk strength. It is also noted that the Weibull modulus for the local strength is much higher than that for bulk strength of glass due to smaller deviation of the local strength.

The results suggested that Hertzian indentation is a potential method for nondestructive evaluation of mechanical properties of brittle materials, especially in those cases listed below.

1. Evaluating the strength and residual stress of equipment made of a brittle material, when cutting this equipment into standard specimens or damaging the material is not an option.
2. Evaluating the mechanical properties of a material made with new technological process, or comparing the properties of two materials.
3. Proof testing of the brittle products.

Acknowledgements

This study was supported by the National Science Foundation of China (No. 59872035, 59902007) and the Science Foundation of Beijing (No. 2002019) and the National Outstanding Young Scientist Foundation (No. 50125204).

References

1. Lawn, B.R., Wilshaw, T.R. (1975) J. Mater. Sci., **10**, 1049
2. Lawn, B.R. (1993) Fracture of Brittle Solid, Cambridge University Press, Cambridge, U.K.
3. Lee, S.K., Lee, K.S., Lawn, B.R., Kim, D.K. (1998) J. Am. Ceram. Soc. **81**, 2061
4. Cook, R.F., Pharr, G.M. (1990) J. Am. Ceram. Soc., **73**, 787
5. Peterson, I.M., Pajares, A., Lawn, B.R., Thompson, V.P., Rekow, E.D. (1998) J. Dent. Res., **77**, 589
6. Robert, S.G., Lawrence, C.W., Bisrat, Y., Warren, P.D., Hills, D.A. (1999) J. Am. Ceram. Soc., **82**, 1809
7. Auerbach, F. (1891) Ann. Phys. Chem., **43**, 61
8. Tillett, J.P. (1956) Proc. Phys. Soc. Lon., **B69**, 981
9. Timoshenko, S., Goodier, J.N. (1951) Theory of Elasticity, McGraw Hill, New York
10. Johnson, K.L. (1985) Contact Mechanics, Cambridge University Press, London, U.K
11. Lawn, B.R. (1988) J. Am. Ceram. Soc., **81**, 1977
12. Bao, Y., Jin, Z. (1993) Fatigue Fract. Engng. Mater. Struct., **16**, 829

14 Machinable Ceramic Graded Composites

W. Pan

14.1 Introduction

Machining is an inevitable requirement for the manufacturing of advanced ceramics, especially for structural ceramics. Current methods for producing ceramic parts rely on finish machining using diamond creep feed grinding or some other non-traditional machining method. However, the high hardness of ceramics makes conventional machining very difficult or even impossible. As a result, machining may represent as much as 90% of the cost of some ceramic parts. In the past years, a lot of research has been focused on the improvement of ceramic machinability [1—10].

14.2 Basic Design Principles

A new materials design methodology for machinable ceramics using graded structure was proposed. Such machinable graded ceramics can combine high-performance with excellent machinability, which is attributed to their unique macrostructure and microstructure design. Fig. 14.1 shows the concept of machinable graded ceramics and a schematic illustration of machinable Si_3N_4/h-BN FGM. For Si_3N_4/h-BN FGM, the graded material is composed of silicon nitride inside and hexagonal boron nitride outside, or silicon nitride at the center and Si_3N_4/h-BN composites at one side or two sides.

The methodology is conceptually fairly simple:

- a layered material added to the conventional advanced ceramic;
- a symmetric gradient distribution and gradient distribution in one direction;
- a process of powder layering.

The main reasons for choosing this gradient design were: first, layered materials have excellent machinability; second, for the Si_3N_4-h-BN and Al_2O_3-$LaPO_4$ systems, there is some degree of similarity between their thermodynamic properties; third, graded design is a good method to eliminate the interface stress of a dissimilar materials system. Based on the thermodynamics and sintering techniques, a variety of machinable graded ceramics can be designed and constructed.

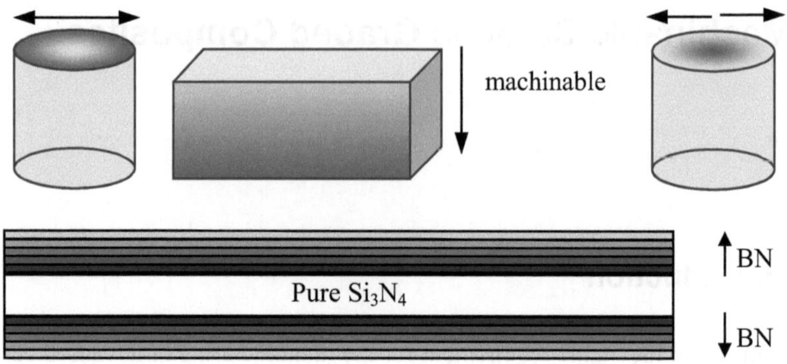

Fig. 14.1. Schematic illustration of machinable Si_3N_4/h-BN FGM

14.3 Machinable Graded Si_3N_4/h-BN Composite

Many methods have been utilized to improve the machinability of Si_3N_4 ceramics, including porous ceramics and microstructure design (controlling the ratio of α-Si_3N_4/β-Si_3N_4). In order to combine the good machinability of h-BN and the excellent mechanical properties of Si_3N_4 at both room temperatures and high temperatures, a graded Si_3N_4/h-BN ceramic was developed in this work. In general, machinability is inversely proportional to hardness and strength. It can be seen that the hardness and bending strength of the Si_3N_4/h-BN composite decreases gradually with increasing h-BN content. The hardness of Si_3N_4/h-BN with 25% h-BN is as high as H_V=5.67 GPa, which matches the needs of machining. This value is close to that of machinable mica glass-ceramic (3 GPa) and layered ternary compounds Ti_3SiC_2 (4~5 GPa).

Figure 14.2 shows the cross-section of the symmetric graded materials composing 11 layers. The layer boundaries are faintly visible in these micrographs. Note the uniformity of the layers in all cases, and the uniformity of the microstructures within individual layers. Most importantly, note the lack of delamination or severe defects at the interlayer boundaries. The higher magnification micrographs of pure Si_3N_4 (middle layer) showed good densification.

Fig. 14.3 shows a hole in the machinable Si_3N_4/h-BN FGM. The hole was produced using a cemented carbide drill. The materials removal mechanism of Si_3N_4/h-BN composite during drilling seems to rely on the cleavage of layered boron nitride crystals. During drilling, the cleavage of h-BN crystals induces fracture of the composite on a microscopic scale and allows chips to form easily, giving rise to good machinability.

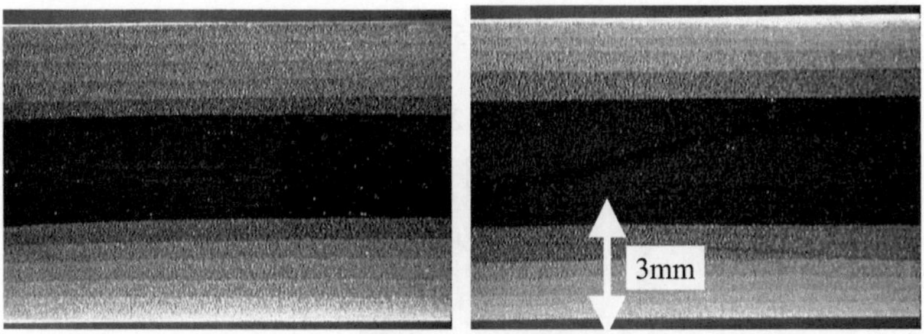

Fig. 14.2. Optical images of machinable Si_3N_4/h-BN FGM

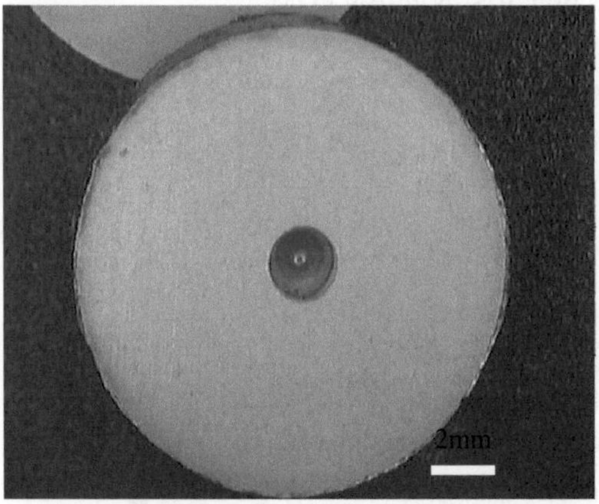

Fig. 14.3. An optical micrograph to show a hole in the machinable Si_3N_4/h-BN FGM

Figure 14.4 illustrates the whole profile of Si_3N_4/h-BN FGM. The figure shows that the porosity of each layer increases with the increase of the h-BN addition. The inter-granular porosity can be seen as another reason for the improved machinability, as it prevents the growth of cracks associated with machining. It seems that the higher the porosity, the better the machinability.

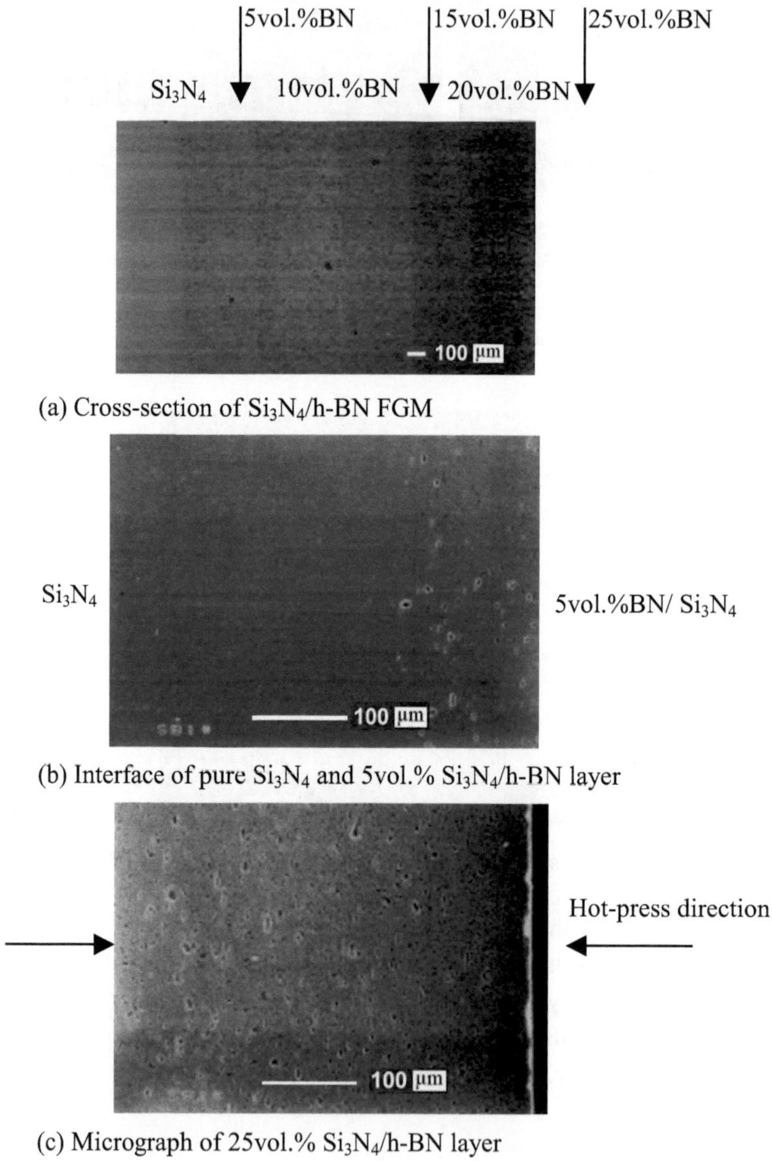

Fig. 14.4. The microstructure of the cross-section of Si_3N_4/h-BN FGM

14.4 Machinable Graded $Al_2O_3/LaPO_4$ Composite

Having different materials and properties at different regions in the same component is desirable for several applications [11]. However, because of the large difference in their coefficients of thermal expansion, the joining of dissimilar materials will lead to the development of stresses, which cause failure. This problem can be solved using the concept of functionally graded materials (FGM). In a FGM, there is a continuous change in composition from one region to another. The FGM can therefore accommodate the shrinkage strain and realize the gradient change of properties. In the present study, an FGM with excellent machinability is explored. In this section, an example of $Al_2O_3/LaPO_4$ FGM is examined.

Figure 14.5 shows the SEM micrographs of fracture sufaces in the symmetrical graded $Al_2O_3/LaPO_4$ composites. According to the figure, the layered $LaPO_4$ crystals tend to align with their easy fracture planes nearly perpendicular to the hot-pressing direction (mechanical aligning). Since the cracks can propagate along the weak interface between Al_2O_3 and $LaPO_4$ phase and layer interface of $LaPO_4$, the addition of layered soft $LaPO_4$ phase can effectively improve the machinability of Al_2O_3 ceramic. The middle layer of the $Al_2O_3/LaPO_4$ FGM is pure Al_2O_3; the $LaPO_4$ content is gradually increased from the center to two sides. As the result of the increased $LaPO_4$ content, composites can be partly machined using cemented carbide drills. The pure Al_2O_3 layer was about 1 mm. Each gradient layer was approximately 0.2 mm thick. With the increased content of $LaPO_4$, the grain size of Al_2O_3 decreased and the grain became easily peeled or flaked due to the weak bonding between Al_2O_3 and $LaPO_4$.

Chemical analysis was performed on a cross-section of the sample in order to evaluate the gradient of $Al_2O_3/LaPO_4$ FGM from pure Al_2O_3 layer to 40wt% $Al_2O_3/LaPO_4$ layer. Fig. 14.7 shows the map and line distribution of elements Al, La and P in the thickness direction of $Al_2O_3/LaPO_4$ FGM. In the cross-section, the La and P contents increase towards the outer surface, whereas the Al content decreases accordingly.

Since layered $LaPO_4$ can be easily delaminated due to its low cleavage energy, fractures propagated parallel to the weak crystallographic direction and the interface between two phases. The $LaPO_4$ addition can induce crack deflection, branching and blunting, which can prevent major cracks from propagating beyond the local cutting area. The weak bonding between $LaPO_4$ and Al_2O_3 can play a similar role during machining.

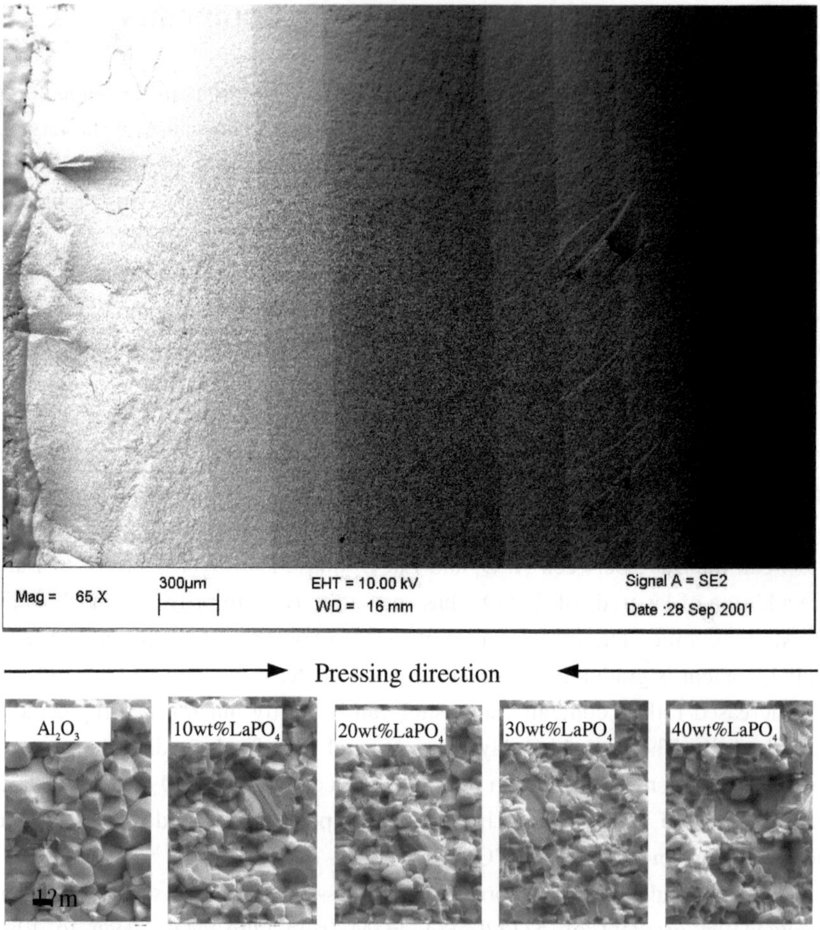

Fig. 14.5. SEM micrographs of fracture surfaces in the graded machinable $Al_2O_3/LaPO_4$ composite

Some parts of graded materials can be easily machined with cemented carbide drills instead of conventional diamond tools as shown in Fig. 14.7. The weak bonding between Al_2O_3 and $LaPO_4$ grains and weak layered structure of $LaPO_4$ grains can blunt the crack tip. Such a role can be interpreted as an energy-loss mechanism. These weak boundaries can absorb or release the fracture energy at the tip of cracks. Considering the macrostructure of $Al_2O_3/LaPO_4$ FGM, the density, elastic modulus, strength, etc. were also gradually changed. For example, due to the layered structure of $LaPO_4$, it is expected that such FGM can also exhibit excellent thermal shock resistance.

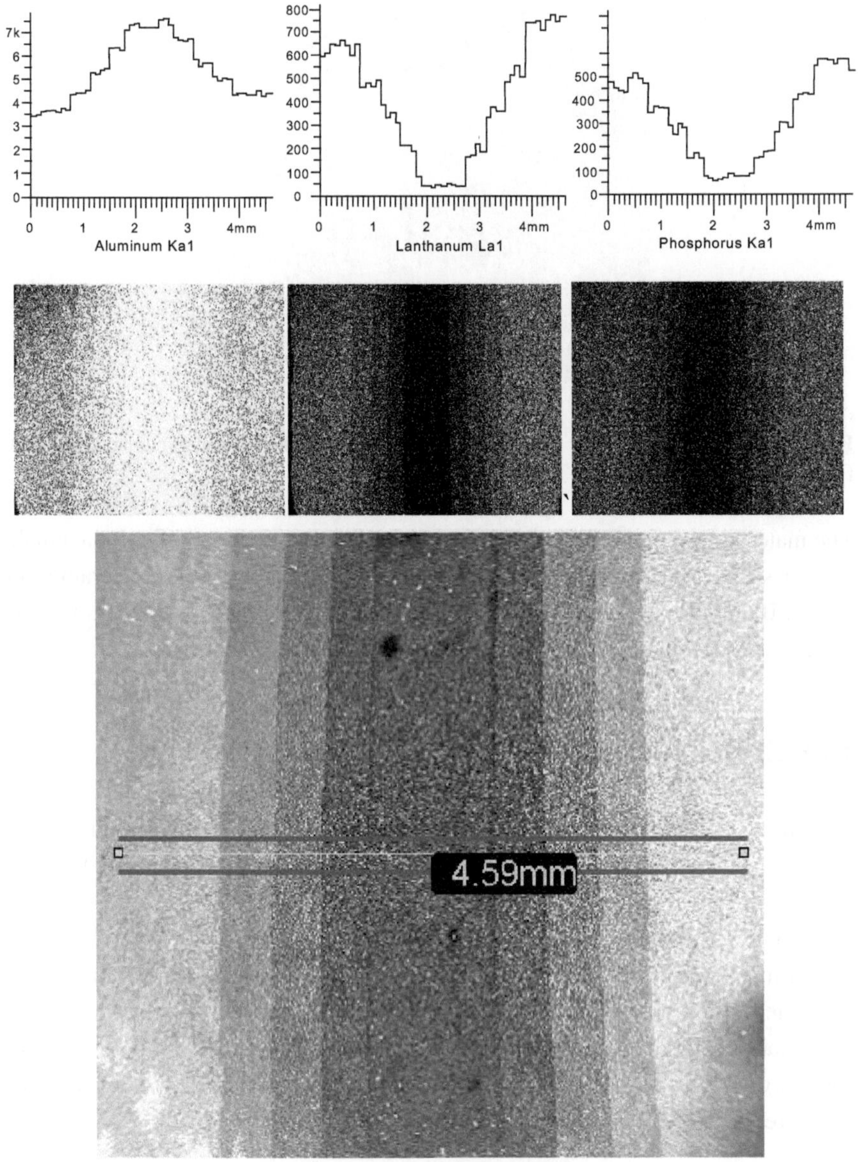

Fig. 14.6. The map and line distribution of Al, P and La concentrations across 1.5 mm of the symmetrical Al_2O_3/$LaPO_4$ FGM

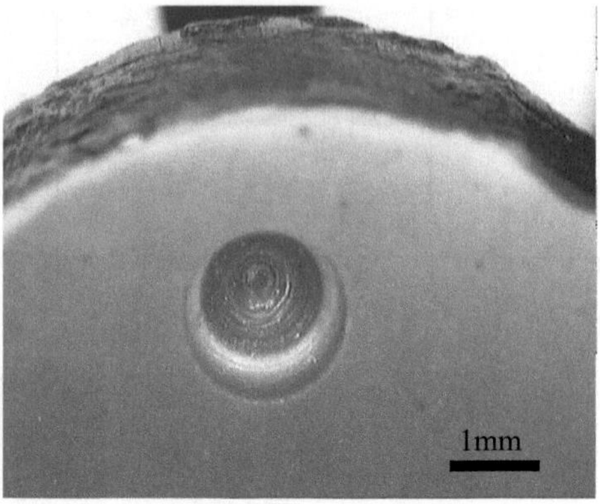

Fig. 14.7. A hole produced by using a cemented carbide drill on a machinable $Al_2O_3/LaPO_4$ FGM

The material design concept of FGM was applied to the fabrication of machinable ceramics in order to develop new machinable and high performance ceramic materials. By employing a graded structure in the design of machinable ceramics, we hope to effectively improve the machinability of advanced ceramics.

14.5 Conclusions

Machining of ceramics can be an expensive processing step, especially when components have a very complex shape and are expected to be machined out of a block or strip. Brittle materials are very sensitive to flaws or cracks, as expressed by their low K_{IC}-values. For a really tough material one would expect the formation of many cracks running in all possible directions without breaking the sample.

In the present study, a functional gradient methodology was applied to design a machinable ceramic. These ceramics can be easily machined using a cemented carbide drill. Furthermore, due to the addition of ceramic crystals with a layered structure, these FGM ceramics are also flaw tolerant.

References

1. Grossman, D.G. (1972) J. Am. Ceram. Soc., **55**, 446
2. Barsoum, M.W., El-Raghy, T. (1996) J. Am. Ceram. Soc., **79**, 1953
3. Padture, N.P., Evans, C.J., Xu, H.H.K. (1995) J. Am. Ceram. Soc., **78**, 215

4. Kawai, C., Yamakawa, A. (1998) J. Ceram. Soc. Jap., **106**, 1135
5. Xu, H.H.K., Jahanmir, S. (1995) , J. Am. Ceram. Soc., **78**, 497
6. Keiichiro, I., Toshihiko, F., Ryoji, U. (1990) Kawasaki Steel Technical Report, **23**, 5
7. Pan, W., Wang, R.G., Jiang, M. N., Liu, F. (2000) Rare Metal Materials and Engineering, **29**, 84
8. Wang, R.G., Pan, W., Jiang, M.N., Chen, J., Luo, Y.M. (2002) Materials Science and Engineering B, **90**, 261
9. Fair, G., Shemkunas, M., Petuskey, W.T. (1999) J. Euro. Ceram. Soc., **19**, 2437
10. Paine, R.T., Narula, C.K. (1990) Chem. Rev., **90**, 73
11. Ruys, A.J., Popoy, E.B., Sun, D., Russell, J.J., Murray, C.C.J. (2001) J. European Ceramic Society, **21**, 2025
12. Pender, D.C., Padture, N.P., Giannakopoulos, A.E., Suresh, S. (2001) Acta Materialia, **49**, 3255
13. Pender, D.C., Thompson, S.C., Padture, N.P., Giannakopoulos, A.E., Suresh, S. (2001) Acta Materialia, **49**, 3263
14. Wang, X.R., Masumoto, H., Someno, Y., Hirai, T. (1998) J. Japan Institute of Metals, **62**,1069
15. Jeon, J.H., Hahn, Y.D., Kim, H.D. (2001) J. European Ceramic Society, **21**, 1653

15 Future of Structural Multi-phase Ceramics

W.-H. Tuan

15.1 Obstacles for the Applications of Ceramics

Ceramics have many attractive features as structural components. To name a few hardness, melting point, chemical inertness, etc. are the best among all materials. However, external stresses are bound to exert on the structural components during application. The poor toughness of ceramics renders their resistance weak to tensile stress. The applications of ceramics as structural components are thus limited by their brittleness.

To improve the toughness of ceramics, many alternatives have been proposed in the last 20 years. One of the most attractive alternatives is to develop ceramic matrix composites. For this approach, a second phase with various shapes is incorporated into the brittle ceramic matrix. However, this approach inevitably increases cost. The applications of these toughened ceramics as structural components are thus limited. Furthermore, the market for structural ceramics is small compared to the sectors for other materials. Without the support of an expanding market, reducing the cost of ceramics seems a rather remote possibility.

What can we ceramists do now? The present author would think the development of better ceramics is always the obligation for ceramists. The market usually responds generously to a breakthrough in the development of new material. For example, the enthusiasm displayed on the discovery of high-temperature superconductors was absolutely overwhelming 10 years ago, though it turned out impractical for many applications in the end.

Therefore, novel materials with one or more outstanding properties are always welcome to the relatively quiet market for ceramics. In the scientific community of late, several experimental evidences are worth noting.

15.2 Novel Materials

These experimental evidences are given in the following. These data are not exhaustive but it givesa sense of the development of new ceramics.

a. High strength multi-phase ceramics [1]. The strength of an Al_2O_3-ZrO_2-SiC composite can reach 1 GPa. The size of SiC powder is around 100 nm, so this is a nano-composite. The total amount of second phase, ZrO_2+SiC, is less than 20 wt%.

b. High strength multi-phase ceramics [2]. The strength of 1000 MPa is reported for an Al_2O_3/t-ZrO_2/m-ZrO_2 composite. The amount of the ZrO_2 is less than 20 vol.%.

c. High toughness multi-phase ceramics [3]. The toughness of an Al_2O_3-ZrO_2-Ni composite can reach 10 $MPam^{0.5}$. The total amount of the second phase is about 25 vol.%. The composite was prepared by pressureless sintering.

d. Superplastic a multi-phase ceramics at elevated temperature [3]. The strain rate of a 30%Al_2O_3-40%ZrO_2-30%$MgAl_2O_4$ composite at high temperature can reach the order of 10^{-2} to 10^{-1} s^{-1}. The composite can show a superplastic deformation of 1050% after stretching at 1650°C for 25 seconds. This high strain rate is close to that of some metal alloys. The common routines for the fabrication of metallic components, such as deep drawing, fast forging, etc. are thus also possible for the manufacture of ceramic components.

15.3 Challenges

Monolithic ceramics have been under extensive research and development for 40 years already. Considerable research resources have been put into the development of monolithic ceramics. Such enthusiasm has generated the so-called "ceramic fever" of the early 1980's. However, the fever faded quickly immediately after the full characterization of monolithic ceramics was done. However, the unique features of ceramics are still there for us to take advantage of. Apart from the features we have known for many years, more unique functions, such as superconductivity, were discovered in the last 20 years. This is not the time to give up all efforts. In the light of limited resources available, the money has to be spent wisely.

The experimental evidence demonstrated in the last section show that there is much space for the improvement of ceramics. The developmemt of multi-phase ceramics seems to be a promising area. Efforts are still needed to demonstrate the feasibility of this approach. However, with the significant improvement in the performance of new ceramics, the ceramic fever (coined by Japanese researchers in 1985) will come back again.

References

1. Guo, J.K. (2001) 2nd Cross-Strait Conference on Ceramic-matrix Composites, Taichung
2. Tuan, W.H., Chen, R.Z., Wang, T.C., Cheng, C.H., Kuo, P.S. (2002) J. European Ceram. Soc., **22**, 2827
3. Chen, R.Z., Chiu, Y.T., Tuan, W.H. (2000) J. European Ceram. Soc., **20**, 1901
4. Kim, B.N., Hiraga, K., Morita, K., Sakka, Y. (2001) Nature **413**, 288

Index

Springer Series in
MATERIALS SCIENCE

Editors: R. Hull R. M. Osgood, Jr. J. Parisi H. Warlimont

10 **Computer Simulation of Ion-Solid Interactions**
By W. Eckstein

11 **Mechanisms of High Temperature Superconductivity**
Editors: H. Kamimura and A. Oshiyama

12 **Dislocation Dynamics and Plasticity**
By T. Suzuki, S. Takeuchi, and H. Yoshinaga

13 **Semiconductor Silicon**
Materials Science and Technology
Editors: G. Harbeke and M. J. Schulz

14 **Graphite Intercalation Compounds I**
Structure and Dynamics
Editors: H. Zabel and S. A. Solin

15 **Crystal Chemistry of High-T_c Superconducting Copper Oxides**
By B. Raveau, C. Michel, M. Hervieu, and D. Groult

16 **Hydrogen in Semiconductors**
By S. J. Pearton, M. Stavola, and J. W. Corbett

17 **Ordering at Surfaces and Interfaces**
Editors: A. Yoshimori, T. Shinjo, and H. Watanabe

18 **Graphite Intercalation Compounds II**
Editors: S. A. Solin and H. Zabel

19 **Laser-Assisted Microtechnology**
By S. M. Metev and V. P. Veiko
2nd Edition

20 **Microcluster Physics**
By S. Sugano and H. Koizumi
2nd Edition

21 **The Metal-Hydrogen System**
By Y. Fukai

22 **Ion Implantation in Diamond, Graphite and Related Materials**
By M. S. Dresselhaus and R. Kalish

23 **The Real Structure of High-T_c Superconductors**
Editor: V. Sh. Shekhtman

24 **Metal Impurities in Silicon-Device Fabrication**
By K. Graff 2nd Edition

25 **Optical Properties of Metal Clusters**
By U. Kreibig and M. Vollmer

26 **Gas Source Molecular Beam Epitaxy**
Growth and Properties of Phosphorus Containing III–V Heterostructures
By M. B. Panish and H. Temkin

27 **Physics of New Materials**
Editor: F. E. Fujita 2nd Edition

28 **Laser Ablation**
Principles and Applications
Editor: J. C. Miller

29 **Elements of Rapid Solidification**
Fundamentals and Applications
Editor: M. A. Otooni

30 **Process Technology for Semiconductor Lasers**
Crystal Growth and Microprocesses
By K. Iga and S. Kinoshita

31 **Nanostructures and Quantum Effects**
By H. Sakaki and H. Noge

32 **Nitride Semiconductors and Devices**
By H. Morkoç

33 **Supercarbon**
Synthesis, Properties and Applications
Editors: S. Yoshimura and R. P. H. Chang

34 **Computational Materials Design**
Editor: T. Saito

35 **Macromolecular Science and Engineering**
New Aspects
Editor: Y. Tanabe

36 **Ceramics**
Mechanical Properties, Failure Behaviour, Materials Selection
By D. Munz and T. Fett

37 **Technology and Applications of Amorphous Silicon**
Editor: R. A. Street

38 **Fullerene Polymers and Fullerene Polymer Composites**
Editors: P. C. Eklund and A. M. Rao

Springer Series in
MATERIALS SCIENCE

Editors: R. Hull R. M. Osgood, Jr. J. Parisi H. Warlimont

39 **Semiconducting Silicides**
Editor: V. E. Borisenko

40 **Reference Materials in Analytical Chemistry**
A Guide for Selection and Use
Editor: A. Zschunke

41 **Organic Electronic Materials**
Conjugated Polymers and Low Molecular Weight Organic Solids
Editors: R. Farchioni and G. Grosso

42 **Raman Scattering in Materials Science**
Editors: W. H. Weber and R. Merlin

43 **The Atomistic Nature of Crystal Growth**
By B. Mutaftschiev

44 **Thermodynamic Basis of Crystal Growth**
P–T–X Phase Equilibrium and Non-Stoichiometry
By J. Greenberg

45 **Thermoelectrics**
Basic Principles and New Materials Developments
By G. S. Nolas, J. Sharp, and H. J. Goldsmid

46 **Fundamental Aspects of Silicon Oxidation**
Editor: Y. J. Chabal

47 **Disorder and Order in Strongly Nonstoichiometric Compounds**
Transition Metal Carbides, Nitrides and Oxides
By A. I. Gusev, A. A. Rempel, and A. J. Magerl

48 **The Glass Transition**
Relaxation Dynamics in Liquids and Disordered Materials
By E. Donth

49 **Alkali Halides**
A Handbook of Physical Properties
By D. B. Sirdeshmukh, L. Sirdeshmukh, and K. G. Subhadra

50 **High-Resolution Imaging and Spectrometry of Materials**
Editors: F. Ernst and M. Rühle

51 **Point Defects in Semiconductors and Insulators**
Determination of Atomic and Electronic Structure from Paramagnetic Hyperfine Interactions
By J.-M. Spaeth and H. Overhof

52 **Polymer Films with Embedded Metal Nanoparticles**
By A. Heilmann

53 **Nanocrystalline Ceramics**
Synthesis and Structure
By M. Winterer

54 **Electronic Structure and Magnetism of Complex Materials**
Editors: D.J. Singh and D. A. Papaconstantopoulos

55 **Quasicrystals**
An Introduction to Structure, Physical Properties and Applications
Editors: J.-B. Suck, M. Schreiber, and P. Häussler

56 **SiO_2 in Si Microdevices**
By M. Itsumi

57 **Radiation Effects in Advanced Semiconductor Materials and Devices**
By C. Claeys and E. Simoen

58 **Functional Thin Films and Functional Materials**
New Concepts and Technologies
Editor: D. Shi

59 **Dielectric Properties of Porous Media**
By S.O. Gladkov

60 **Organic Photovoltaics**
Concepts and Realization
Editors: C. Brabec, V. Dyakonov, J. Parisi and N. Sariciftci

MIX
Papier aus verantwortungsvollen Quellen
Paper from responsible sources
FSC® C105338

If you have any concerns about our products,
you can contact us on
ProductSafety@springernature.com

In case Publisher is established outside the EU,
the EU authorized representative is:
Springer Nature Customer Service Center GmbH
Europaplatz 3, 69115 Heidelberg, Germany

Printed by Libri Plureos GmbH
in Hamburg, Germany